INSTITUTE OF GEOLOGICAL SCIENCES

Natural Environment Research Council

Mineral Assessment Report 52

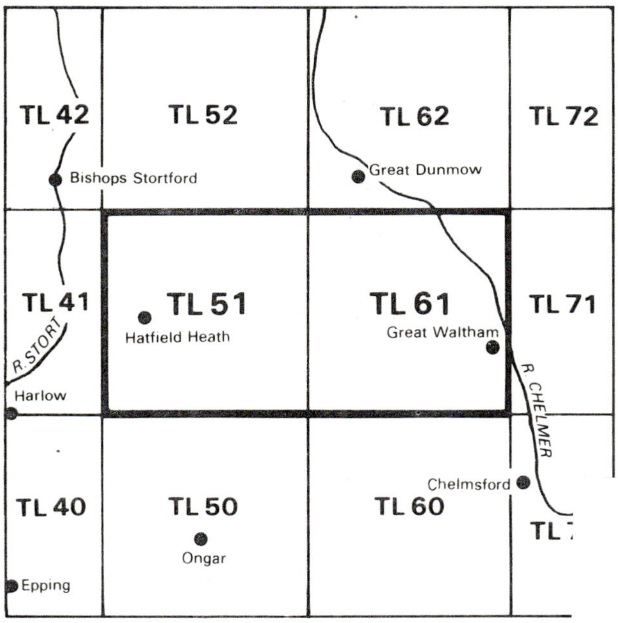

The sand and gravel resources of the country between Hatfield Heath and Great Waltham, Essex

Description of 1:25 000 sheets TL 51/61

ISBN 0 11 884113 0

London Her Majesty's Stationery Office 1980

The first twelve reports on the assessment of British sand and gravel resources appeared in the Report series of the Institute of Geological Sciences as a sub-series. Report 13 and subsequent reports appear as Mineral Assessment Reports of the Institute.

Details of published reports appear at the end of this Report.

Any enquiries concerning this report may be addressed to Head, Industrial Minerals Assessment Unit, Institute of Geological Sciences, Keyworth, Nottingham NG12 5GG.

PREFACE

National resources of many industrial minerals may seem so large that stocktaking appears unnecessary, but the demand for minerals and for land for all purposes is intensifying and it has become increasingly clear in recent years that regional assessments of the resources of these minerals should be undertaken. The publication of information about the quantity and quality of deposits over large areas is intended to provide a comprehensive factual background against which planning decisions can be made.

Sand and gravel, considered together as naturally occurring aggregate, was selected as the bulk mineral demanding the most urgent attention, initially in the south-east of England, where about half the national output is won and very few sources of alternative aggregates are available. Following a short feasibility project, initiated in 1966 by the Ministry of Land and Natural Resources, the Industrial Minerals Assessment Unit, formerly the Mineral Assessment Unit, began systematic surveys in 1968. The work is now being financed by the Department of the Environment and is being undertaken with the cooperation of the Sand and Gravel Association of Great Britain.

This report describes the resources of sand and gravel of 200 km^2 of country between Hatfield Heath and Great Waltham, Essex, shown on the accompanying 1:25 000 resource sheet (TL 51/61). The survey was conducted by Mr R. J. Marks, who was assisted in the drilling and sampling programme by Mr R. W. Gatliff. The work is based on six-inch scale geological surveys by Mr R. A. Ellison, Mr M. J. Heath, Mr R. D. Lake and Dr B. S. P. Moorlock in 1969–1976.

Mr J. W. Gardner, CBE, (Land Agent) was responsible for negotiating access to land for drilling. The ready cooperation of land owners, tenants, Messrs Cawoods Aggregates Limited, and the assistance of officials of the Essex County Council is gratefully acknowledged.

G. M. Brown
Director

Institute of Geological Sciences
Exhibition Road
London SW7 2DE

March 1980

CONTENTS

FIGURES

MAP

TABLES

The sand and gravel resources of the country between Hatfield Heath and Great Waltham, Essex

Description of 1:25 000 sheets TL 51/61

R. J. MARKS

SUMMARY

The geological maps of the Institute of Geological Sciences, pre-existing borehole information, and 129 boreholes drilled for the Industrial Minerals Assessment Unit, form the basis of the assessment of sand and gravel resources in the area between Hatfield Heath and Great Waltham, Essex.

All deposits in the area which might be potentially workable for sand and gravel have been investigated and a simple statistical method has been used to estimate the volume. The reliability of the volume estimates is given at the symmetrical 95 per cent probability level.

The 1:25 000 map is divided into five resource blocks, containing between 6.6 and 12.6 km^2 of potentially workable sand and gravel. For each block the geology of the deposits is described and the mineral-bearing area, the mean thicknesses of overburden and mineral, and the mean gradings are stated. Detailed borehole data are also given. The geology, the position of the boreholes and the outlines of the resource blocks are shown on the accompanying map.

Note

All grid references quoted in this report fall within the 100-km grid square TL.

Bibliographical reference

MARKS, R. J. 1980. The sand and gravel resources of the country between Hatfield Heath and Great Waltham, Essex. Description of 1:25 000 resource sheet TL 51/61. *Miner. Assess. Rep. Inst. Geol. Sci.*, No. 52.

Author

R.J.Marks, BSc
Institute of Geological Sciences
Keyworth
Nottingham NG12 5GG

INTRODUCTION

The survey is concerned with the estimation of resources, which include deposits that are not currently exploitable but have a foreseeable use, rather than reserves, which can only be assessed in the light of current, locally prevailing, economic considerations. Clearly, both the economic and the social factors used to decide whether a deposit may be workable in the future cannot be predicted; they are likely to change with time. Deposits not currently economically workable may be exploited as demand increases, as higher grade or alternative materials become scarce, or as improved processing techniques are applied to them. The improved knowledge of the main physical properties of the resource and their variability which this survey seeks to provide, will add significantly to the factual background against which planning policies can be decided (Archer, 1969; Thurrell, 1971; Harris and others, 1974).

The survey provides information at the 'indicated' level "for which tonnage and grade are computed partly from specific measurements, samples or production data and partly from projection for a reasonable distance on geological evidence. The sites available for inspection, measurement, and sampling are too widely spaced to permit the mineral bodies to be outlined completely or the grade established throughout" (Bureau of Mines and Geological Survey, 1948, p. 15).

It follows that the whereabouts of reserves must still be established and their size and quality proved by the customary detailed exploration and evaluation undertaken by the industry. However, the information provided by this survey should assist in the selection of the best targets for such further work.

The following arbitrary physical criteria have been adopted:

a The deposit should average at least 1 m in thickness.
b The ratio of overburden to sand and gravel should be no more than 3:1.
c The proportion of fines (particles passing the No. 240 mesh BS sieve, about $\frac{1}{16}$ mm) should not exceed 40 per cent.
d The deposit should lie within 25 m of the surface, this being taken as the likely maximum working depth under most circumstances. It follows from the second criterion that boreholes are drilled no deeper than 18 m if no sand and gravel has been proved.

A deposit of sand and gravel which broadly meets these criteria is regarded as 'potentially workable' and is described and assessed as 'mineral' in this report. As the assessment is at the indicated level, parts of such a deposit may not satisfy all the criteria.

For the particular needs of assessing sand and gravel resources, a grain-size classification based on the geometric scale $\frac{1}{16}$ mm, $\frac{1}{4}$ mm, 1 mm, 4 mm, 16 mm has been adopted. The boundaries between fines (that is, the clay

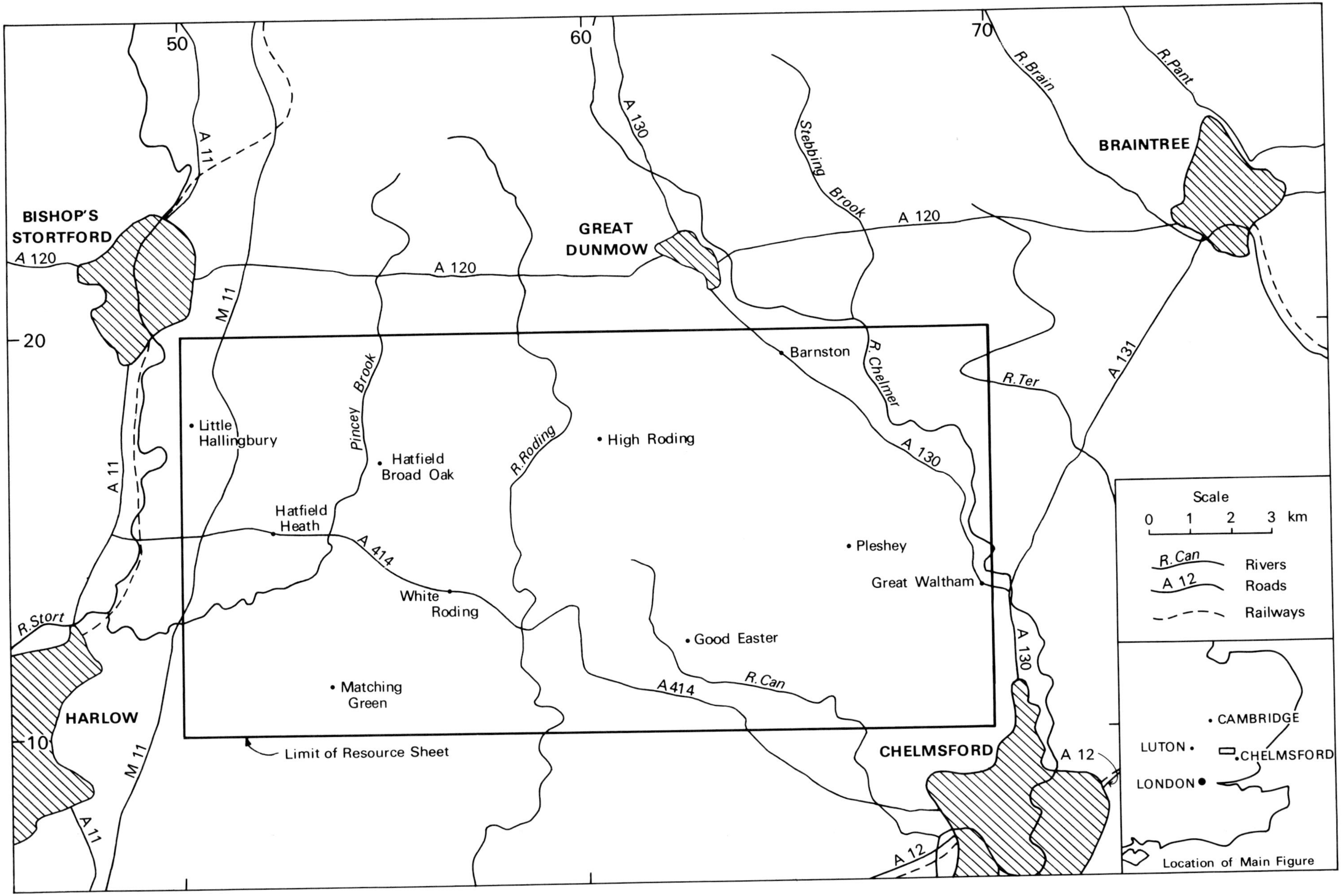

Figure 1 Map showing the location of 1:25 000 sheets TL 51/61

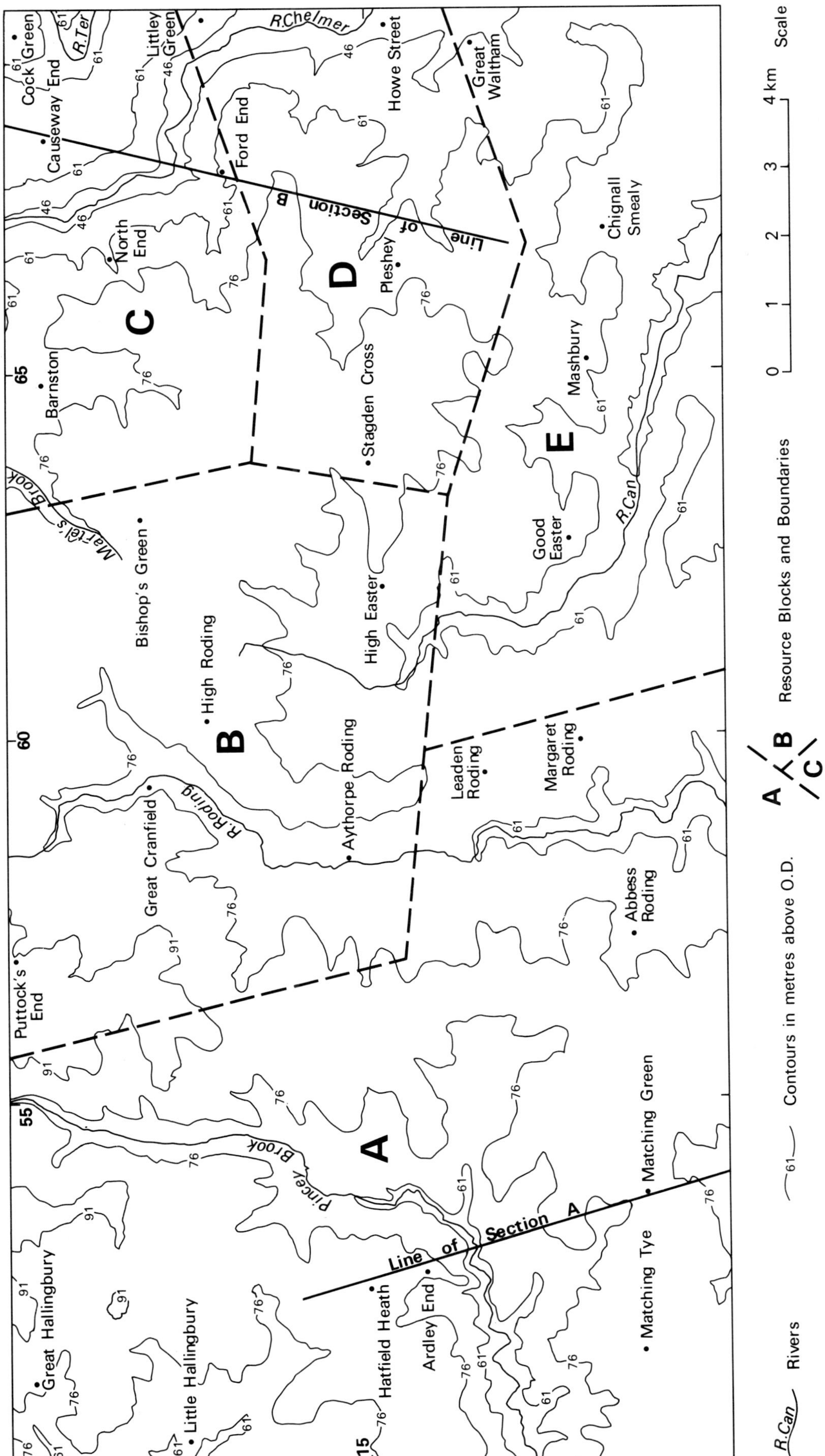

Figure 2 Locality map

and silt fractions) and sand, and between sand and gravel grade material, are placed at $\frac{1}{16}$ mm and 4 mm respectively (see Appendix C).

The volume and other characteristics are assessed within resource blocks, each of which, ideally, contains approximately 10 km^2 of sand and gravel. No account is taken of any factors, for example, roads, villages and high agricultural or landscape value, which might stand in the way of sand and gravel being exploited, although towns are excluded. The estimated total volume therefore bears no simple relationship to the amount that could be extracted in practice.

It must be emphasised that the assessment applies to the resource block as a whole. Valid conclusions cannot be drawn about the mineral in parts of a block, except in the immediate vicinity of the actual sample points.

DESCRIPTION OF THE DISTRICT

GENERAL

The district covers 200 km^2 of country in south-west Essex, west and north of Chelmsford (Figure 1). It supports arable farming and is renowned for the wheat lands of the Roding Valley. This agriculture is based on the heavy soils which are derived from the chalky Boulder Clay that underlies most of the area.

The main topographic feature of the district (Figure 2) is a gently undulating surface, which represents part of the Essex and Suffolk plateau; in this area it slopes from +90 m Ordnance Datum in the north-west to about +60 m Ordnance Datum in the south-east. It is dissected by four generally southward-flowing rivers, the Roding, Can, Chelmer and Ter, as well as Pincey and Martel's brooks.

Although potentially workable sand and gravel underlies 49.1 km^2 it is extracted only in the south-east of the area from two pits [688 122; 659 103] at Margret at Woods near Broad's Green and Chignall Hall.

GEOLOGY

The geology of this area has been described in a Geological Survey memoir (Whitaker and others, 1878). The geological sequence is summarised in Table 1 and the relationships between the deposits are illustrated in the sections of Figure 3; the lines of section are shown in Figure 2 and on the Map.

Table 1 Geological sequence

		Maximum recorded thickness m
DRIFT		
Recent and	Calcareous Tufa	*
Pleistocene	Alluvium	8.1
	Head	4.2
	Head Gravel	6.5
	Boulder Clay	27.4
	Glacial Sand and Gravel	10.7
SOLID		
Eocene	London Clay	42.1
Palaeocene	Lower London Tertiaries	8.8+

* deposit not sampled
\+ base not reached

SOLID

The extensive cover of thick superficial deposits restricts exposures of solid rocks in the area mainly to the Chelmer and Ter valleys in the north-east; the presence of solid formations elsewhere is known from boreholes.

Lower London Tertiaries: The Lower London Tertiaries are described by Hester (1965). These rocks occur in the north-west part of the district and have a low regional dip to the south-east. They comprise two formations: the Thanet Beds and the Woolwich and Reading Beds, which are not easily distinguished in this area because they are not separated by a defined marker horizon.

Borehole 51 NW 53 [5038 1969] encountered these rocks and proved, beneath Head, 1.5 m of mottled clay with sand laminae on 6.3 m of yellow-orange clayey sand, comprising the Woolwich and Reading Beds, and 1.0 m of mottled green and grey clayey sand, interpreted as Thanet Beds. The full thickness of Thanet Beds in this area was not proved.

London Clay: Apart from a small area in the north-west, London Clay forms the bedrock over the whole of the district and rests upon the Lower London Tertiaries. London Clay crops out in the Chelmer Valley, its tributary valleys around Barnston [649 195] and in the Can Valley [616 161] north of High Easter.

Assessment boreholes 51 NW 54 [5024 1838], 51 NW 55 [5066 1737] and 51 NW 56 [5034 1676] proved a sequence of bluish black to olive-grey, silty sands and clays, typical of the lower part of the London Clay, and the more familiar lithology of homogeneous, stiff, olive-grey clay was proved in boreholes elsewhere. This clay is often micaceous and sporadically contains selenite (calcium sulphate) crystals and cementstones. Palaeontological remains include shell fragments, wood and burrows, some of which are pyritised. Beneath Drift deposits, the London Clay is weathered brown and orange-brown to a depth of about 0.4 m; in contrast, where the clay is exposed the weathering penetrates to about 5 m.

DRIFT

Some refinements in the divisions of the Drift have recently been proposed by Rose, Allen and Hey (1976) and Rose and Allen (1977). The Drift consists largely of Boulder Clay and Glacial Sand and Gravel, which form a thick sequence over most of the district: it rests on the eroded surface (rockhead) of the solid formations (Figure 4). The main feature of this surface is an east–west depression, which has been recognised by other workers and termed the 'mid-Essex' depression by Wooldridge and Henderson (1955). It contrasts markedly with the present topography (Figure 2), except in the north-east part of the district, where the Chelmer Valley has been cut through the Drift deposits into the bedrock surface. At White Roding, [563 136] in contrast, an estimated 40 m of Drift separates the plateau surface from the depression beneath.

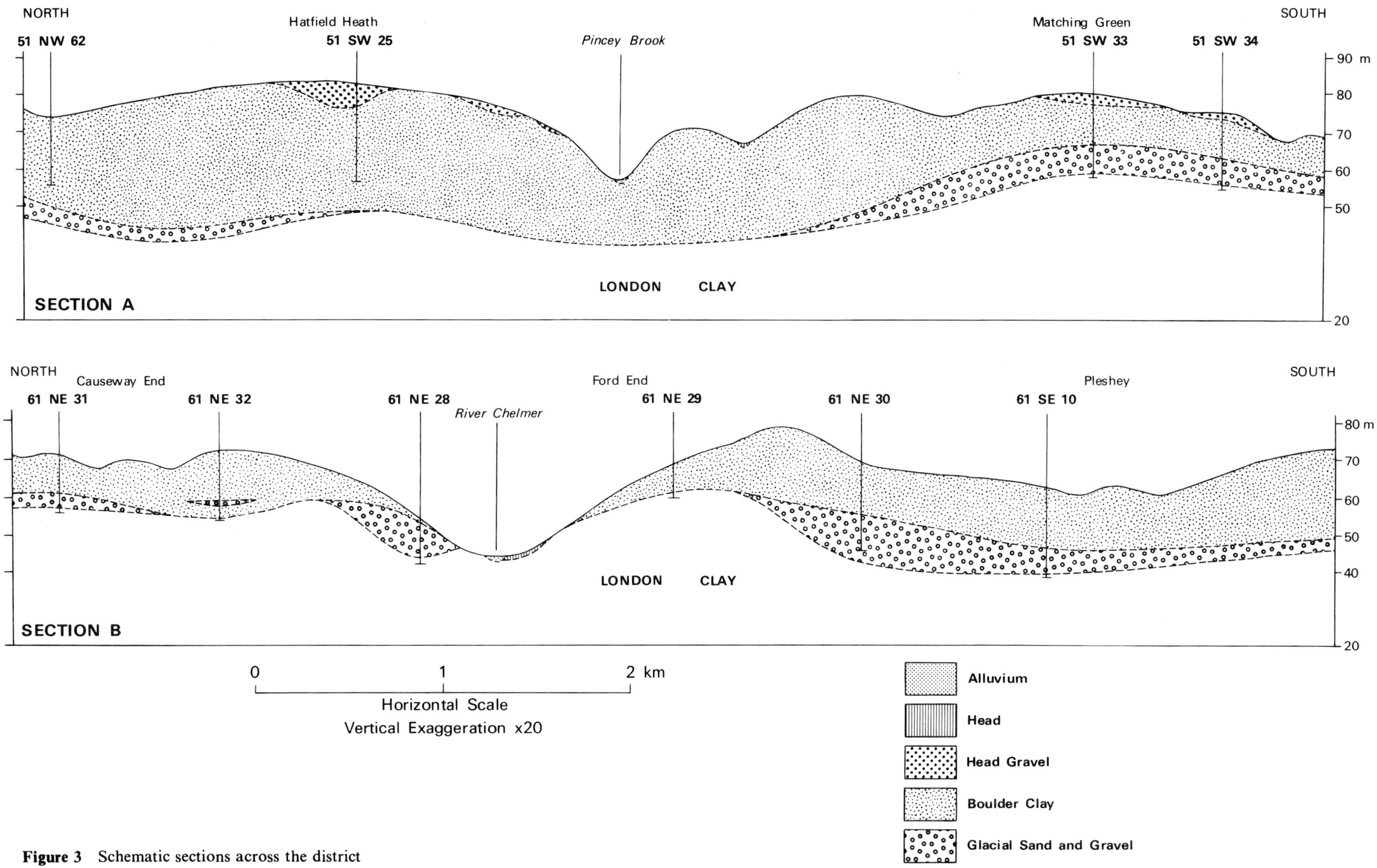

Figure 3 Schematic sections across the district

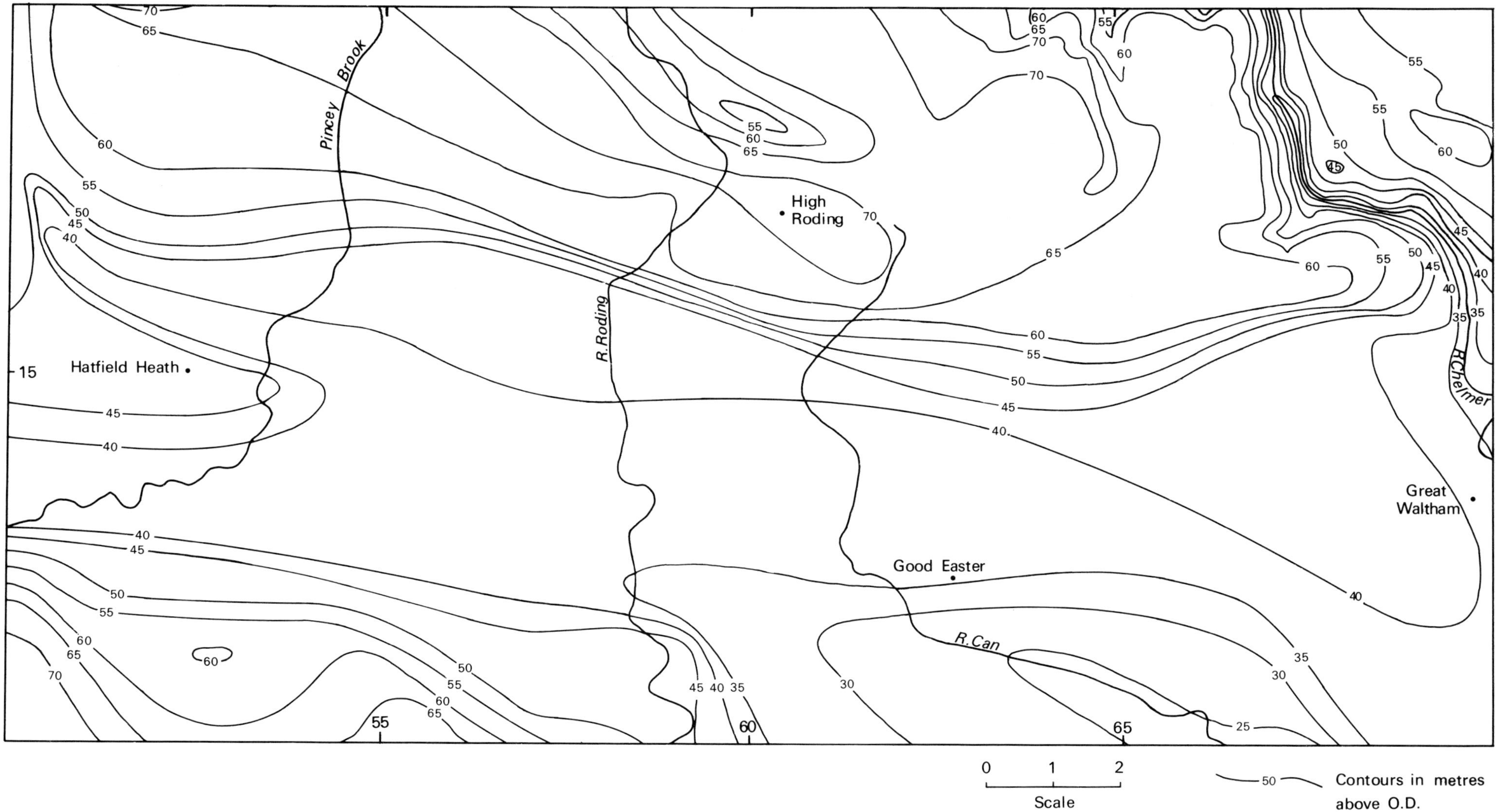

Figure 4 Map showing rockhead contours

Glacial Sand and Gravel: These deposits are similar to those encountered during the assessment of other areas in Essex, for example around Terling and Harlow (Eaton, 1973; Hopson, 1979), and are similarly interpreted as the product of a braided stream environment prior to the main glacial advance. Borehole records suggest that the Glacial Sand and Gravel rests on the subdrift surface as a discontinuous sheet beneath a generally thick sequence of Boulder Clay, while a few (for example, borehole 61 NE 32 [6825 1895]) prove Glacial Sand and Gravel intercalations within the Boulder Clay (Figure 3). Except in the workings at Margret at Woods and Chignall Hall, exposures of the Glacial Sand and Gravel are generally restricted to the more deeply incised valleys.

The deposits reach a maximum recorded thickness of 10.7 m in borehole 61 NE 28 [6789 1793] and generally prove to be well-sorted clean gravel composed largely of flint with subordinate amounts of quartzite and quartz.

Exposures reveal bedding structures of cross sets and channels depicted by varying proportions of sand and gravel and these are well displayed in the gravel pits at Margret at Woods and Chignall Hall.

Boulder Clay: An almost continuous sheet of Boulder Clay covers the area. It underlies more recent sediments on the plateau and in the valleys, and is absent in the deeply dissected parts of the Chelmer, Roding and Ter Valleys, where underlying deposits are revealed.

A tripartite lithological sequence is evident in the deposit. A basal unit of olive-brown, pebbly, sandy clay is developed locally. The pebbles are of well-rounded flint, rounded quartz and quartzite with angular flint and a trace of fine chalk. This is overlain by a thicker unit of olive-black, stiff, slightly sandy, pebbly clay, whose pebbles are largely fine in grade and include shale, flint and some chalk. Above occurs the chalky Boulder Clay, the thickest unit. It is a grey clay with many pebbles of rounded chalk and some angular flint, with shale pebbles present towards the base. This uppermost unit includes seams of orange-grey laminated clay, 'clayey' sand, chalk gravel, and sand and gravel. Typically the basal unit is absent and the Boulder Clay has a sharp junction with the underlying deposits, while the junctions between the units are gradational. The Boulder Clay has a maximum recorded thickness of 27.4 m at borehole 51 SE 46 [5725 1146]. This sequence of Boulder Clay is interpreted as the deposit of a single glacial advance.

This deposit is modified by a weathering profile, which penetrates to an average depth of 4 m. Beneath the soil is a brown, slightly sandy clay containing some angular flints; with depth, many chalk pebbles are present and the clay becomes mottled brown and light grey. Towards the base of the weathering profile the brown mottling fades and the clay takes on a brown-grey colour before becoming grey and unweathered.

Head Gravel: In the west of the district, between Little Hallingbury [503 174] and Matching Green [536 111], this deposit rests on Boulder Clay in a sinuous tract of isolated bodies with a maximum recorded thickness of 6.5 m in borehole 51 SW 25 [5258 1497]. The deposit is poorly sorted and includes a range of lithologies from 'clayey' gravel to pebbly, sandy clay; it is mottled orange-brown and light grey near the surface and becomes orange-brown with depth. The pebble content is typically of angular flint with subordinate quartzite and quartz; exceptionally borehole 51 SW 25 has a high proportion of chalk. The Head Gravel is thought, due to its stratigraphic position and structureless form, to have been deposited as the ice retreated.

Head: This solifluction deposit blankets the lower slopes and floors of the valleys and rests on older Drift and Solid formations. It consists of unsorted locally-derived clay, silt, sand and pebbles.

Alluvium: Alluvium occurs as a continuous veneer along the lower reaches of the valley floors where it rests on older Drift and Solid formations. It consists of a layered sequence, laid down by the river of each valley. The basal unit which is not always present is a gravel composed largely of angular flint pebbles with sporadic clay and silt seams; it is well developed in the Roding Valley. A middle unit of silty peat is developed commonly and contains plant debris and fresh-water gastropod shells, while the top unit consists of silty, sandy clay with sporadic flint pebbles.

Calcareous Tufa: A small deposit of Calcareous Tufa [638 197] is present south-east of Martel's Brook. It is a soft, light grey silt-grade deposit formed of calcium carbonate, which has been held in solution by percolating water. The calcium carbonate is thought to have been derived from chalk in the Boulder Clay.

COMPOSITION OF SAND AND GRAVEL

Glacial Sand and Gravel: The potentially workable Glacial Sand and Gravel has a mean grading of fines 7 per cent, sand 45 per cent and gravel 48 per cent (that is gravel, see Figure 8), and ranges to 'very clayey' sandy gravel around Great Canfield [594 180] and North End [667 185].

Coarse gravel (+16 mm) predominates over fine (+4−16 mm) in the west of the area, whereas the reverse is true in the east; cobbles are virtually absent everywhere. The major component of the gravel fraction is angular to subangular flint (Table 2). Subordinate lithologies include subrounded to rounded quartzite, well rounded flint, subrounded to rounded quartz and a trace of subangular to subrounded sandstone, subangular ironstone, rounded metasediments and igneous rocks.

The sand fraction has an average grading of fine sand 19 per cent, medium sand 59 per cent and coarse sand 22 per cent; it ranges from angular to rounded and is composed of quartz with subordinate comminuted flint.

Borehole samples from this deposit show that the fines content decreases with depth, with the gravel fraction showing a corresponding increase; typically, the basal metre shows an increased percentage of sand. Silt and clay seams are developed sporadically throughout the deposit.

Head Gravel: This potentially workable deposit is generally 'clayey' gravel with a mean grading of fines 16 per cent, sand 38 per cent and gravel 46 per cent. The gravel typically contains both fine and coarse fractions; it is composed largely of angular to subangular flint with subordinate subrounded to rounded quartzite, well-rounded flint, subrounded to rounded quartz and a trace of sandstone, ironstone, metasediments and igneous rocks (Table 2). Exceptionally however, borehole 51 SW 25 proved that the gravel contained 19 per cent chalk and 4 per cent ironstone—constituents that would be regarded as deleterious in many applications.

Table 2 Pebble count analyses

Borehole number	Percentage by weight						
	Angular* flint	Well-rounded flint	Quartzite	Quartz	Sandstone	Chalk	Others
ALLUVIAL GRAVEL							
51 NE 14	55	21	10	7	2	2	3
51 NE 17	62	11	17	4	2	0	4
51 SE 39	59	19	15	3	2	0	2
51 SE 40	65	8	14	7	0	4	2
51 SE 44	65	10	9	5	0	6	5
Mean	61	14	13	5	1	3	3
HEAD GRAVEL							
51 SW 25	59	4	5	2	4	19	7
51 SW 26	65	6	20	6	1	0	2
51 SW 27	62	11	16	8	1	0	2
51 SW 29	71	8	12	4	2	0	3
51 SW 33	72	6	12	6	0	0	4
51 SW 34	78	8	8	2	3	0	1
Mean	68	7	12	5	2	3	3
GLACIAL SAND AND GRAVEL							
51 NW 54	52	5	20	15	3	0	5
51 NW 61	53	13	23	8	1	0	2
51 NE 6	41	17	19	19	1	0	3
51 NE 12	51	16	25	4	2	0	2
51 NE 14	57	15	19	6	1	0	2
51 NE 16	44	16	21	14	2	0	3
51 NE 18	30	22	37	7	2	0	2
51 SW 33	42	24	25	6	1	0	2
51 SW 34	52	14	16	15	0	0	3
51 SW 37	45	19	23	12	0	0	1
51 SE 36	51	17	17	12	0	0	3
51 SE 40	56	35	5	3	0	0	1
51 SE 41	51	25	16	5	0	0	3
51 SE 44	49	44	4	2	0	0	1
61 NW 8	42	18	26	10	0	0	4
61 NW 9	58	10	18	8	4	0	2
61 NW 12	35	25	33	4	2	0	1
61 NE 18	48	20	19	9	3	0	1
61 NE 19	45	21	23	6	3	0	2
61 NE 21	45	26	21	4	1	0	3
61 NE 24	39	28	24	7	1	0	1
61 NE 28	44	19	24	9	1	0	3
61 NE 31	48	16	25	9	0	0	2
61 NE 34	42	19	32	3	3	0	1
61 NE 55	46	18	22	9	1	0	4
61 SW 15	72	8	17	2	1	0	0
61 SE 6	50	18	22	8	0	0	2
61 SE 11	55	19	18	5	1	0	2
61 SE 13	57	18	15	7	1	0	2
61 SE 18	51	22	21	4	2	0	0
61 SE 19	55	13	23	6	2	0	1
61 SE 21	65	15	14	4	2	0	0
Mean	49	19	21	8	1	0	2

* Includes a range from angular to subangular flint.

The mean grading of the sand fraction is fine sand 18 per cent, medium sand 51 per cent and coarse sand 31 per cent; it is predominantly composed of quartz with subordinate flint.

Alluvium: The mineral-bearing Alluvium is confined to the Roding Valley and is typically 'clayey' gravel with a mean grading of fines 15 per cent, sand 35 per cent and gravel 50 per cent. The gravel fraction is predominantly of coarse grade and composed largely of angular flint with subordinate well-rounded flint, subrounded quartzite and quartz with a trace of chalk, sandstone, metasediments and igneous rocks.

The sand fraction has a mean grading of fine sand 21 per cent, medium sand 49 per cent and coarse sand 30 per cent, and is composed of angular to subrounded quartz with subordinate angular to subangular flint and a trace of subrounded to rounded chalk.

THE MAP

The sand and gravel resource map is folded into the pocket at the end of this report. The base map is the Ordnance Survey 1:25 000 Outline Edition in grey, on which the topography is shown by contours in green, the geological data in black and the mineral resource information in shades of red.

Table 3 Summary of statistical results

Resource block	Area		Mean thickness		Volume of mineral			Mean grading percentage		
						Limits at the 95% probability level				
	Block km^2	Mineral km^2	Over-burden m	Mineral m	million m^3	$\pm$%	$\pm$million m^3	Fines $-\frac{1}{16}$ mm	Sand $+\frac{1}{16}-4$ mm	Gravel $+4$ mm
A	79.6	11.3	8.1	5.1	58	28	16	9	40	51
B	42.3	12.6	9.7	5.1	64	34	22	10	49	41
C	21.4	6.6	3.9	5.0	33	52	17	11	45	44
D	23.0	10.4	11.1	6.4	67	31	21	6	46	48
E	33.7	8.2	7.7	6.4	52	33	17	5	37	58
Total	200.0	49.1	8.4	5.5	270	13	35	9	44	47

Geological data: The geological boundary lines and symbols are taken from the geological maps of the area, which was surveyed recently at the scale of 1:10 560 by Field Staff of the Institute's East Anglia and South-East England Unit. Borehole data, which include the stratigraphic relations, thicknesses and mean particle-size distribution of the sand and gravel samples collected during the assessment survey, are also shown.

The geological boundaries are the best interpretation of the information available at the time of survey. However, it is inevitable, particularly with Drift deposits, which change rapidly vertically and laterally, that local discrepancies may occur.

Mineral resource information: The mineral-bearing ground is subdivided into resource blocks (see Appendix A). Within a resource block the mineral is subdivided into areas where it is 'exposed' and areas where it is present beneath overburden. The mineral is identified as 'exposed' where the overburden, commonly consisting only of soil and subsoil, averages less than 1.0 m in thickness. Beneath overburden the mineral may be continuous (or almost continuous) or discontinuous. The recognition of these categories is dependent upon the importance attached to the proportion of boreholes which did not find potentially workable sand and gravel and the distribution of barren boreholes within a block. The mineral is described as 'almost continuous' if it is present in 75 per cent or more of the boreholes in a resource block. The 'discontinuous' category has not been recognised on the present sheet.

Areas where bedrock crops out and where the available evidence suggests that sand and gravel is not potentially workable or is absent, are uncoloured on the Map; where appropriate the reason is given. In such areas it has been assumed that mineral is absent except in infrequent and relatively minor patches, which can neither be outlined nor assessed quantitatively in the context of this survey. Areas of unassessed sand and gravel are indicated by a red stipple.

The area of the exposed sand and gravel is measured from the mapped geological boundary lines. The whole of this area is considered as mineral, although it may include small areas where sand and gravel is not present or is not potentially workable. Inferred boundaries have been inserted where sand and gravel beneath cover is interpreted to be not potentially workable or absent. Such boundaries (for which a distinctive symbol is used) are drawn primarily for the purpose of volume estimation. The symbol is intended to convey an approximate location within a likely zone of occurrence rather than to represent the breadth of the zone, its size being limited only by cartographic considerations. For the purpose of measuring areas the centre-line of the symbol is used.

RESULTS

The statistical results are summarised in Table 3. Fuller grading particulars are shown in Figure 5, and Tables 4 to 8.

For the five resource blocks, the confidence limits at the symmetrical 95 per cent probability level vary between 28 per cent and 52 per cent (that is, it is probable that nineteen times out of twenty the true volumes present lie within these limits). However, the true values are more likely to be nearer the figures estimated than the limits. Moreover, it is probable that in each block roughly the same percentage limits would apply for the estimate of volume of a very much smaller parcel of ground (say, 100 hectares) containing similar sand and gravel deposits if results from the same number of sample points (as provided by, say, ten boreholes) were used in the calculation. Thus, if closer limits are needed for the quotation of reserves of part of a block, it can be expected that data from more than ten sample points will be required, even if the area is quite small. This point can be illustrated by considering the whole of the potentially workable sand and gravel on this sheet. The volume (270 million m^3) can be estimated to limits of ± 13 per cent at the 95 per cent probability level, by a calculation based on the data from 55 sample points in the five resource blocks.

However, it must be emphasised that the quoted volume of sand and gravel has no simple relationship with the amount that could be extracted in practice, as no allowance has been made in the calculations for any restraints (such as existing buildings and roads) on the use of the land for mineral working.

NOTES ON RESOURCE BLOCKS

The district is divided into five resource blocks (Table 3). The boundaries between resource blocks mainly reflect the distribution of the principal resource, the Glacial Sand and Gravel. The extent of this resource depends largely on the thickness of overburden, which becomes excessive (p. 1) over the central part of the district, coinciding with the 'mid-Essex depression' (Figure 4) where the drift deposits are thickest. The other potentially workable deposits are Head Gravel, which is confined to block A, and the Alluvium of the Roding Valley, which is divided between blocks A and B.

Table 4 Block A: data from assessment boreholes

Borehole number	Recorded thickness		Mean grading percentage					
	Mineral m	Over-burden m	Fines $-\frac{1}{16}$ mm	Fine sand $+\frac{1}{16}-\frac{1}{4}$ mm	Medium sand $+\frac{1}{4}-1$ mm	Coarse sand +1–4 mm	Fine gravel +4–16 mm	Coarse gravel +16 mm
ALLUVIUM								
51 SE 40	1.1	1.6	13	6	30	15	25	11
51 SE 44	2.0	1.2	12	3	15	10	34	26
Mean	1.5	1.4	12	4	23	12	30	19
HEAD GRAVEL								
51 NW 60	Mineral absent							
51 SW 25	6.0	0.5	13	5	15	16	23	28
51 SW 26	3.6	0.5	22	13	24	8	16	17
51 SW 27	1.2	0.1	13	4	13	12	30	28
51 SW 29	3.8	0.6	16	5	18	13	28	20
51 SW 33	2.8	0.1	14	8	26	12	22	18
51 SW 34	1.2	0.5	16	6	21	10	21	26
Mean	2.9	0.4	16	7	19	12	23	23
GLACIAL SAND AND GRAVEL								
51 NW 53	Mineral absent							
51 SW 33	6.8*	10.7	3	3	29	11	25	29
51 SW 34	2.6*	11.1	2	10	23	7	29	29
51 SW 37	6.2	16.0	6	17	17	8	20	32
51 SW 38	7.1	10.8	3	8	26	8	25	30
51 SE 36	6.7	15.1	4	5	17	10	29	35
51 SE 37	5.8	15.6	8	7	25	14	18	28
51 SE 40	3.0	3.6	11	31	36	5	14	13
51 SE 41	7.9	6.6	2	7	19	9	28	35
51 SE 44	3.9	4.7	4	5	21	9	28	33
Mean	5.4	11.0	4	10	24	9	24	29

* Thickness excludes waste

Block A

This block covers 79.6 km² of country between Great Hallingbury [512 196], Matching Tye [515 113] and Margaret Roding [599 120]; it encloses 11.3 km² of mineral, which includes Glacial Sand and Gravel, Head Gravel and Alluvium (Table 4).

The Glacial Sand and Gravel is potentially workable around Great Hallingbury, Matching Green and to the south-east of Abbess Roding [572 114], while the intervening areas are classified as non-mineral, due primarily to an excessive Boulder Clay thickness. Additionally, borehole 51 NW 61 [5244 1706] proved 7.5 m of buried mineral but the area of the occurrence cannot be defined at the scale of this survey and it has not been included in the assessment. Ten IMAU boreholes and seven other boreholes have been used to assess the potentially workable Glacial Sand and Gravel, which has a maximum recorded thickness of 7.9 m in borehole 51 SE 41 [5824 1048], a mean thickness of 5.4 m, and covers an area of 9.0 km². The deposit grades as gravel at all sample points except borehole 51 SE 40 [5888 1118], where it is sandy gravel. A waste seam comprising pebbly sandy clay splits the mineral in boreholes 51 SW 33 [5350 1115] and 51 SW 34 [5385 1052]. Apart from three small outcrops north of Little Hallingbury, the mineral is beneath overburden, mostly Boulder Clay, which ranges in recorded thickness up to 16 m with a mean of 11 m.

Head Gravel occurs as discrete patches resting upon Boulder Clay between Little Hallingbury and Matching Green. The assessment is based on seven IMAU and two other boreholes, with boreholes 51 NW 60 [5180 1529] and 51 SW 28 [5232 1217] proving non-mineral sequences of pebbly sandy clay. Potentially workable Head Gravel covers 2.9 km², ranges in recorded thickness from 1.2 m to 6.0 m with a mean of 2.9 m, and grades as 'clayey' gravel. At Matching Green (boreholes 51 SW 33 and 51 SW 34) resources of Head Gravel are separated from the underlying mineral-bearing Glacial Sand and Gravel by 10.6 m of Boulder Clay.

Head Gravel has been worked locally at Hatfield Heath [524 150], south-west of Ardley End [527 143] and Waterman's End [538 105]. Because of their small size several isolated areas of Head Gravel have been left unassessed, for example near Shell House [5400 1984], Greenhill [553 186] and Old Street Hill [535 158].

The Alluvium of the Roding Valley includes a thin basal sand and gravel, which is estimated to become potentially workable south of Pig's Bridge [5852 1159], where it exceeds 1 m thick. This mineral deposit rests on Boulder Clay, which overlies potentially workable Glacial Sand and Gravel. The mineral covers 0.3 km² and two IMAU boreholes (51 SE 40 [5888 1118], 51 SE 44 [5925 1035]) proved a mean thickness of 1.5 m of 'clayey' gravel overlain by Alluvium with a mean thickness of 1.4 m (Table 4).

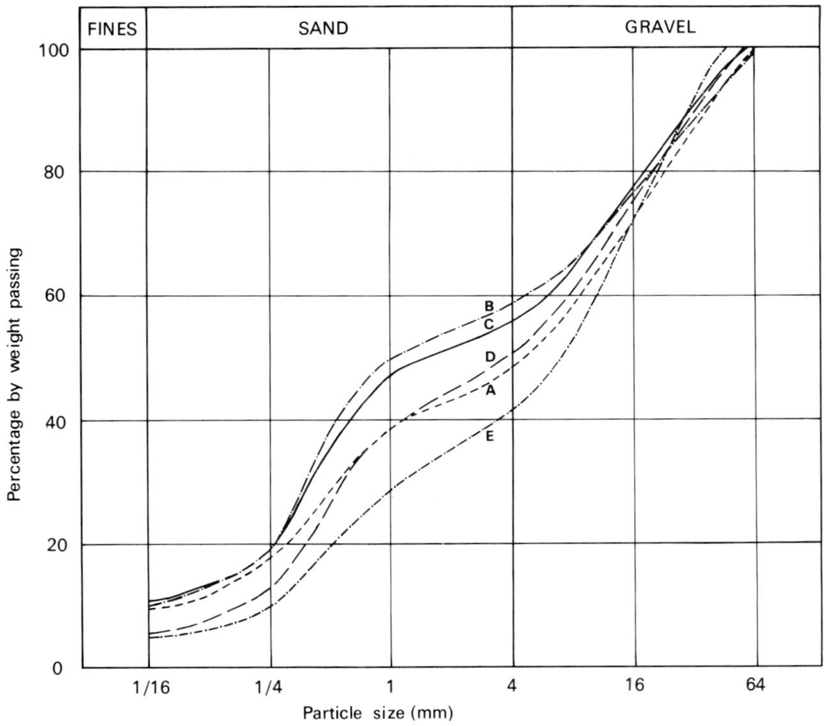

Resource Block	Percentage by weight passing				
	1/16mm	1/4mm	1mm	4mm	16mm
A	9	18	39	49	73
B	10	19	50	59	77
C	11	19	47	56	78
D	6	13	39	52	76
E	5	10	29	42	73

Figure 5 Mean particle-size distribution for the assessed thickness of sand and gravel in the resource blocks

The combined mineral of the block grades as gravel with a mean grading of fines 9 per cent, sand 40 per cent and gravel 51 per cent, with an estimated volume of 58 million $m^3 \pm 28$ per cent at the 95 per cent probability level.

Block B
This block, which encompasses Puttock's End [569 198], Aythorpe Roding [583 152], High Easter [622 148] and Bishop's Green [630 179] covers 42.3 km^2 and includes 12.6 km^2 of mineral in the Glacial Sand and Gravel and Alluvium (Table 5).

Potentially workable Glacial Sand and Gravel is present under the plateau areas of Puttock's End, and between High Roding [603 173] and Bishop's Green, as well as along the Roding Valley. Ten IMAU boreholes proved mineral which ranges in recorded thickness between 2.8 m and 8.7 m, with a mean of 5.2 m, and grades as sandy gravel. Though there are small outcrops of Glacial Sand and Gravel along the valley of the River Roding, between Bloomshawbury [574 168] and Black Hall [582 184] and along Martel's Brook, most of the resource is covered by overburden, largely Boulder Clay, which ranges in recorded thickness up to 16.1 m with a mean of 10.4 m.

Sand and gravel at the base of the Alluvium in the north and south of the block is too thin to be regarded as mineral, but in the intervening area of the Roding Valley, boreholes 51 NE 14 [5812 1639] and 51 NE 17 [5927 1845] proved 1 m and 2.5 m of 'clayey' gravel respectively beneath Alluvium overburden with a mean thickness of 2.1 m. This deposit is largely underlain by mineral-bearing Glacial Sand and Gravel, which appears to be reworked in borehole 51 NE 17 to form the basal gravel sequence of the Alluvium.

The combined potentially workable deposits of this block grade as sandy gravel with a mean grading of fines 10 per cent, sand 49 per cent and gravel 41 per cent, with an estimated volume of 64 million $m^3 \pm 34$ per cent.

Table 5 Block B: data from assessment boreholes

Borehole number	Recorded thickness		Mean grading percentage					
	Mineral m	Over-burden m	Fines $-\frac{1}{16}$ mm	Fine sand $+\frac{1}{16}-\frac{1}{4}$ mm	Medium sand $+\frac{1}{4}-1$ mm	Coarse sand $+1-4$ mm	Fine gravel $+4-16$ mm	Coarse gravel $+16$ mm
ALLUVIUM								
51 NE 14	1.0	1.3	15	5	15	11	30	24
51 NE 17	2.5	2.9	20	15	10	6	23	26
Mean	1.8	2.1	18	10	12	8	27	25
GLACIAL SAND AND GRAVEL								
51 NE 6	8.7	13.8	6	6	25	15	25	23
51 NE 14	8.3	12.5	10	21	41	6	12	10
51 NE 16	7.2	11.8	14	8	38	8	16	16
51 NE 18	2.8	4.3	2	2	30	9	25	32
61 NW 4	4.5	16.1	22	13	39	6	10	10
61 NW 5	3.2	14.7	4	4	21	6	20	45
61 NW 8	6.3	11.5	6	4	32	10	21	27
61 NW 9	3.0	5.1	14	5	44	11	11	15
61 NW 11	2.8	5.5	2	3	21	14	27	33
61 NW 12	5.5	8.8	15	21	45	4	7	8
Mean	5.2	10.4	9	9	34	9	17	22

Block C

With an area of 21.4 km^2 this block straddles the Chelmer Valley between Barnston, Cock Green [696 198] and Ford End [677 168]. The only potentially workable deposit is Glacial Sand and Gravel, which covers 6.6 km^2. Except for small outcrops of mineral where the river valleys are incised into it, the deposit is present beneath overburden of Boulder Clay which, however, is relatively thin locally, and areas of mineral beneath overburden have been identified: they may be more or less extensive than shown on the map. Exceptionally, a thin bed of Glacial Sand and Gravel was proved by borehole 61 NE 32 within the Boulder Clay. Several small outcrops of Glacial Sand and Gravel in the Chelmer Valley appear, from borehole evidence, to be of only local significance and are therefore unassessed.

The assessment is based on six IMAU boreholes and one other; they indicate (Table 6) a maximum recorded thickness of 10.7 m of mineral with a mean of 5 m. This block is estimated to contain 33 million m^3 $\pm$ 52 per cent of mineral, which grades as 'clayey' sandy gravel. Proved overburden thicknesses range up to 10.2 m with a mean of only 3.9 m.

The deposit formerly worked at Causeway End [683 193] was proved in borehole 61 NE 32 to be pebbly and sandy Boulder Clay with a fines content in excess of 40 per cent; it is therefore considered as non-mineral.

Table 6 Block C: data from assessment boreholes

Borehole number	Recorded thickness		Mean grading percentage					
	Mineral m	Over-burden m	Fines $-\frac{1}{16}$ mm	Fine sand $+\frac{1}{16}-\frac{1}{4}$ mm	Medium sand $+\frac{1}{4}-1$ mm	Coarse sand $+1-4$ mm	Fine gravel $+4-16$ mm	Coarse gravel $+16$ mm
GLACIAL SAND AND GRAVEL								
61 NE 18	4.0	5.0	8	9	31	8	23	21
61 NE 19	4.6	5.9	10	5	29	9	24	23
61 NE 23	4.2	0.4	21	15	23	6	18	17
61 NE 24	5.1	1.9	13	5	27	12	20	23
61 NE 28	10.7	0.2	10	5	27	11	26	21
61 NE 31	3.9	10.2	4	6	33	7	21	29
Mean	5.0	3.9	11	8	28	9	22	22

Table 7 Block D: data from assessment boreholes

Borehole number	Recorded thickness		Mean grading percentage					
	Mineral m	Overburden m	Fines $-\frac{1}{16}$ mm	Fine sand $+\frac{1}{16}-\frac{1}{4}$ mm	Medium sand $+\frac{1}{4}-1$ mm	Coarse sand $+1-4$ mm	Fine gravel $+4-16$ mm	Coarse gravel $+16$ mm
GLACIAL SAND AND GRAVEL								
61 NE 21	5.7	15.9	4	3	26	12	24	31
61 NE 30	9.8+	14.2	8	12	34	11	20	15
61 NE 34	3.2	7.8	9	7	33	13	20	18
61 NE 35	8.4	13.1	3	3	21	16	27	30
61 SE 6	9.2	13.2	3	5	22	13	29	28
61 SE 10	7.0	16.2	2	4	27	13	26	28
61 SE 11	5.6	16.7	4	3	16	12	32	33
61 SE 15	6.3	8.6	11	18	38	9	15	9
61 SE 18	7.4	3.3	7	10	20	10	27	26
Mean	6.4	11.1	6	7	26	12	25	24

+ sign indicates that the full thickness of mineral was not proved in the borehole.

Block D

Potentially workable Glacial Sand and Gravel extends over 10.4 km² of this block, which stretches between Stagden Cross [637 149], Littley Green [698 171] and Howe Street [697 147], covering an area of 23 km². To the north-east, this area abuts on the Chelmer Valley towards which the overburden thins. To the west and south however, the block is underlain by the northern margin of the 'mid-Essex depression' (Figure 4) where the Boulder Clay thickness exceeds 18 m, the maximum depth to which boreholes are taken in overburden.

Based on the results of nine IMAU boreholes and one other borehole the mineral in this block has a mean thickness of 6.4 m with a maximum recorded thickness of 9.8 m in borehole 61 NE 30 [6742 1564]. It grades as gravel (Table 7) with an estimated volume of 67 million $m^3 \pm 31$ per cent (Table 3).

The overburden is composed largely of Boulder Clay and has a mean thickness of 11.1 m and ranges up to 16.7 m in borehole 61 SE 11 [6754 1368].

In the north, the Chelmer Valley has cut through the Boulder Clay and the underlying Glacial Sand and Gravel, which outcrops in isolated areas on the valley sides; it has been extracted in the past at Fitzandrew's Farm [6944 1484].

Block E

This block covers 33.7 km² east of Leaden Roding [595 133] and includes the parishes of Good Easter, Mashbury, Chignall and Great Waltham. In this block 8·2 km² of Glacial Sand and Gravel is potentially workable and it occurs almost exclusively beneath overburden of Boulder Clay. In the west the sand and gravel is too deeply buried to be regarded as potentially workable, but in the east the overburden is thinner and the sand and gravel is mineral.

The assessment of resources is based on 3 IMAU boreholes (Table 8) and other records; the mineral has a mean thickness of 6.4 m and recorded thicknesses range from 1.2 m to 9.9 m. The estimated volume of mineral is 52 million $m^3 \pm 33$ per cent, which grades as gravel (Table 3).

Overburden ranges up to 16.3 m in recorded thickness and has a mean value of 7.7 m. Glacial Sand and Gravel is currently being extracted from two pits, at Margret at Woods and Chignall Hall, and it is estimated that 4 million m³ of sand and gravel has so far been extracted.

Table 8 Block E: data from assessment boreholes

Borehole number	Recorded thickness		Mean grading percentage					
	Mineral m	Overburden m	Fines $-\frac{1}{16}$ mm	Fine sand $+\frac{1}{16}-\frac{1}{4}$ mm	Medium sand $+\frac{1}{4}-1$ mm	Coarse sand $+1-4$ mm	Fine gravel $+4-16$ mm	Coarse gravel $+16$ mm
GLACIAL SAND AND GRAVEL								
61 SE 13	8.0	4.6	7	7	24	12	29	21
61 SE 19	6.8	5.0	4	4	15	14	31	32
61 SE 21	9.9	7.7	5	4	18	14	32	27
Mean	6.4	7.7	5	5	19	13	31	27

APPENDIX A

FIELD AND LABORATORY PROCEDURES

Trial and error during initial studies of the complex and variable glacial deposits of East Anglia and Essex showed that an absolute minimum of five sample points evenly distributed across the sand and gravel are needed to provide a worthwhile statistical assessment, but that, where possible, there should be not less than ten. Sample points are any points for which adequate information exists about the nature and thickness of the deposit and may include boreholes other than those drilled during the survey and exposures. In particular, the cooperation of sand and gravel operators ensures that boreholes are not drilled where reliable information is already available; although this may be used in the calculations, it is held confidentially by the Institute and cannot be disclosed.

The mineral shown on each 1:25 000 sheet is divided into resource blocks. The arbitrary size selected is a compromise to meet the aims of the survey by providing sufficient sample points in each block. As far as possible the block boundaries are determined by geological boundaries so that, for example, glacial and river terrace gravels are separated. Otherwise division is by arbitrary lines, which may bear no relationship to the geology. The blocks are drawn provisionally before drilling begins.

A reconnaissance of the ground is carried out to record any exposures and inquiries are made to ascertain what borehole information is available. Borehole sites are then selected to provide an even pattern of sample points at a density of approximately one per square kilometre. However, because broad trends are independently overlain by smaller scale characteristically random variations, it is unnecessary to adhere to a square grid pattern. Thus such factors as ease of access and the need to minimise disturbance to land and the public are taken into account in siting the holes; at the same time it is necessary to guard against the possibility that ease of access (that is, the positions of roads and farms) may reflect particular geological conditions, which may bias the drilling results.

The drilling machine employed should be capable of providing a continuous sample representative of all unconsolidated deposits, so that the in-situ grading can be determined, if necessary, to a depth of 30 m at a diameter of about 200 mm, beneath different types of overburden. It should be reliable, quiet, mobile and relatively small (so that it can be moved to sites of difficult access). Shell and auger rigs have proved to be almost ideal.

The rigs are modified to enable deposits above the water table to be drilled 'dry', instead of with water added to facilitate the drilling, to minimise the amount of material drawn in from outside the limits of the hole. The samples thus obtained are representative of the in-situ grading, and satisfy one of the most important aims of the survey. Below the water table the rigs are used conventionally, although this may result in the loss of some of the fines fraction and the pumping action of the bailer tends to draw unwanted material into the hole from the sides or the bottom.

A continuous series of bulk samples is taken throughout the sand and gravel. Ideally samples are composed exclusively of the whole of the material encountered in the borehole between stated depths. However, care is taken to discard, as far as possible, material which has caved or has been pumped from the bottom of the hole. A new sample is commenced whenever there is an appreciable lithological change within the sand and gravel, or at every 1 m depth. The samples, each weighing between 25 and 45 kg, are despatched in heavy duty polythene bags to a laboratory for grading. The grading procedure is based on British Standard 1377 (1967). Random checks on the accuracy of the grading are made in the Institute's laboratories.

All data, including mean grading analysis figures calculated for the total thickness of the mineral, are entered on standard record sheets, abbreviated copies of which are reproduced in Appendix F.

Detailed records may be consulted at the appropriate offices of the Institute, upon application to the Head, Industrial Minerals Assessment Unit.

APPENDIX B

STATISTICAL PROCEDURE

Statistical assessment

1 A statistical assessment is made of an area of mineral greater than 2 km², if there is a minimum of five evenly spaced boreholes in the resource block (for smaller areas see paragraph 12 below).

2 The simple methods used in the calculations are consistent with the amount of data provided by the survey. Conventional symmetrical confidence limits are calculated for the 95 per cent probability level, that is, there is a 5 per cent or one in twenty chance of a result falling outside the stated limits.

3 The volume estimate (V) for the mineral in a given block is the product of the two variables, the sampled areas (A) and the mean thickness ($\bar{l}_m$) calculated from the individual thicknesses at the sample points. The standard deviations for these variables are related such that

$$S_V = \sqrt{(S_A^2 + S_{\bar{l}_m}^2)} \quad [1]$$

4 The above relationship may be transposed such that

$$S_V = S_{\bar{l}_m}\sqrt{(1 + S_A^2/S_{\bar{l}_m}^2)} \quad [2]$$

From this it can be seen that as $S_A^2/S_{\bar{l}_m}^2$ tends to 0, S_V tends to $S_{\bar{l}_m}$.

If, therefore, the standard deviation for area is small with respect to that for mean thickness, the standard deviation for volume approximates to that for mean thickness.

5 Given that the number of approximately evenly spaced sample points in the sampled area is n with mineral thickness measurements $l_{m_1}, l_{m_2}, \ldots l_{m_n}$, then the best estimate of mean thickness, $\bar{l}_m$, is given by

$$\Sigma(l_{m_1} + l_{m_2} \ldots l_{m_n})/n.$$

For groups of closely spaced boreholes a discretionary weighting factor may be applied to avoid bias (see note on weighting below). The standard deviation for mean thickness $S_{\bar{l}}$, expressed as a proportion of the mean thickness, is given by

$$S_{\bar{l}} = (1/\bar{l}_m)\sqrt{[\Sigma(l_m - \bar{l}_m)^2/(n-1)]}$$

where l_m is any value in the series l_{m_1} to l_{m_n}.

6 The sampled area in each resource block is coloured pink on the map. Wherever possible, calculations relate to the mineral within mapped geological boundaries (which may not necessarily correspond to the limits of deposit). Where the area is not defined by a mapped boundary, that is, where the boundary is inferred, a distinctive symbol is used. Experience suggests that the errors in determining area are small relative to those in thickness. The relationship $S_A/S\bar{l}_m \leqslant \frac{1}{3}$ is assumed in all cases. It follows from equation [2] that

$$S_{\bar{l}_m} \leqslant S_V \leqslant 1.05\, S_{\bar{l}_m} \quad [3]$$

7 The limits on the estimate of mean thickness of mineral, $L_{\bar{l}_m}$, may be expressed in absolute units $\pm(t/\sqrt{n}) \times S_{\bar{l}_m}$ or as a percentage $\pm(t/\sqrt{n}) \times S_{\bar{l}_m} \times (100/\bar{l}_m)$ per cent, where t is Student's t at the 95 per cent probability level for $(n-1)$ degrees of freedom, evaluated by reference to statistical tables. (In applying Student's t it is assumed that the measurements are distributed normally).

Block calculation **1:25 000 Block** Fictitious

Area
Block: 11.08 km^2
Mineral: 8.32 km^2

Mean thickness
Overburden: 2.5 m
Mineral: 6.5 m

Volume
Overburden: 21 million m^3
Mineral: 54 million m^3

Confidence limits of the estimate of mineral volume at the 95 per cent probability level: ± 20 per cent
That is, the volume of mineral (with 95 per cent probability): 54 ± 11 million m^3

Thickness estimate measurements in metres
l_o = overburden thickness l_m = mineral thickness

Sample point	Weighting w	Over-burden l_o	wl_o	Mineral l_m	wl_m	Remarks
SE 14	1	1.5	1.5	9.4	9.4	IMAU boreholes
SE 18	1	3.3	3.3	5.8	5.8	
SE 20	1	nil	–	6.9	6.9	
SE 22	1	0.7	0.7	6.4	6.4	
SE 23	1	6.2	6.2	4.1	4.1	
SE 24	1	4.3	4.3	6.4	6.4	
SE 17	$\frac{1}{2}$	1.2	1.6	9.8	7.2	
123/45	$\frac{1}{2}$	2.0		4.6		Hydrogeology Unit record
1	$\frac{1}{4}$	2.7	2.6	7.3	5.8	Close group of four boreholes (commercial)
2	$\frac{1}{4}$	4.5		3.2		
3	$\frac{1}{4}$	0.4		6.8		
4	$\frac{1}{4}$	2.8		5.9		
Totals	$\Sigma w = 8$	$\Sigma wl_o = 20.2$		$\Sigma wl_m = 52.0$		

Means $\overline{wl_o} = 2.5$ $\overline{wl_m} = 6.5$

Calculation of confidence limits

wl_m	$(wl_m - \overline{wl_m})$	$(wl_m - \overline{wl_m})^2$
9.4	2.9	8.41
5.8	0.7	0.49
6.9	0.4	0.16
6.4	0.1	0.01
4.1	2.4	5.76
6.4	0.1	0.01
7.2	0.7	0.49
5.8	0.7	0.49

$\Sigma(wl_m - \overline{wl_m})^2 = 15.82$
$n = 8$
$t = 2.365$

L_V is calculated as

$$1.05(t/\overline{wl_m})\sqrt{[\Sigma(wl_m - \overline{wl_m})^2/n(n-1)]} \times 100$$
$$= 1.05 \times (2.365/6.5)\sqrt{[15.82/(8 \times 7)]} \times 100$$
$$= 20.3$$
$$\simeq 20 \text{ per cent.}$$

Figure 6 Example of resource block assessment: calculation and results

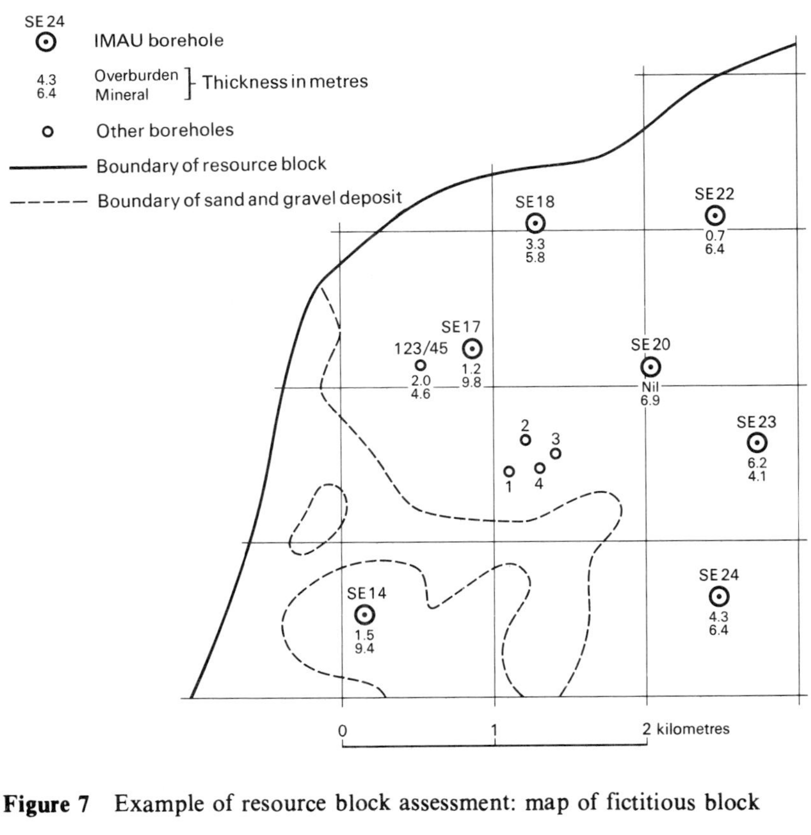

Figure 7 Example of resource block assessment: map of fictitious block

8 Values of t at the 95 per cent probability level for values of n up to 20 are as follows:

n	t	n	t
1	infinity	11	2.228
2	12.706	12	2.201
3	4.303	13	2.179
4	3.182	14	2.160
5	2.776	15	2.145
6	2.571	16	2.131
7	2.447	17	2.120
8	2.365	18	2.110
9	2.306	19	2.101
10	2.262	20	2.093

(from Table 12, Biometrika Tables for Statisticians, Volume 1, Second Edition, Cambridge University Press, 1962). When n is greater than 20, 1.96 is used (the value of t when n is infinity).

9 In calculating confidence limits for volume, L_V, the following inequality corresponding to equation [3] is applied: $L_{\bar{l}_m} \leqslant L_V \leqslant 1.05\, L_{\bar{l}_m}$

10 In summary, for values of n between 5 and 20, L_V is calculated as

$$[(1.05 \times t)/\bar{l}_m] \times [\sqrt{\Sigma(l_m - \bar{l}_m)^2/n(n-1)}] \times 100$$

per cent, and when n is greater than 20, as

$$[(1.05 \times 1.96)/\bar{l}_m] \times [\sqrt{\Sigma(l_m - \bar{l}_m)^2/n(n-1)}] \times 100$$

per cent.

11 The application of this procedure to a fictitious area is illustrated in Figures 6 and 7.

Inferred assessment

12 If the sampled area of mineral in a resource block is between 0.25 km^2 and 2 km^2 an assessment is inferred, based on geological and topographical information usually supported by the data from one or two boreholes. The volume of mineral is calculated as the product of the area, measured from field data, and the estimated thickness. Confidence limits are not calculated.

13 In some cases a resource block may include an area left uncoloured on the map, within which mineral (as defined) is interpreted to be generally absent. If there is reason to believe that some mineral may be present, an inferred assessment may be made.

14 No assessment is attempted for an isolated area of mineral less than 0.25 km^2.

15 *Note on weighting* The thickness of a deposit at any point may be governed solely by the position of the point in relation to a broad trend. However, most sand and gravel deposits also exhibit a random pattern of local, and sometimes considerable, variation in thickness. Thus the distribution of sample points need be only approximately regular and in estimating the mean thickness only simple weighting is necessary. In practice, equal weighting can often be applied to thicknesses at all sample points. If, however, there is a distinctly unequal distribution of points, bias is avoided by dividing the sampled area into broad zones, to each of which a value roughly proportional to its area is assigned. This value is then shared between the data points within the zone as the weighting factor.

APPENDIX C

CLASSIFICATION AND DESCRIPTION OF SAND AND GRAVEL

For the purposes of assessing resources of sand and gravel a classification should take account of economically important characteristics of the deposit, in particular the absolute content of fines and the ratio of sand to gravel.

The terminology commonly used by geologists when describing sedimentary rocks (Wentworth, 1922) is not entirely satisfactory for this purpose. For example, Wentworth proposed that a deposit should be described as a 'gravelly sand' when it contains more sand than gravel and there is at least 10 per cent of gravel, provided that there is less than 10 per cent of material finer than sand (less than $\frac{1}{16}$ mm) and coarser than pebbles (more than 64 mm in diameter). Because deposits containing more than 10 per cent fines are not embraced by this system a modified binary classification based on Willman (1942) has been adopted.

When the fines content exceeds 40 per cent the material is not considered to be potentially workable and falls outside the definition of mineral. Deposits which contain 40 per cent fines or less are classified primarily on the ratio of sand to gravel but qualified in the light of the fines content, as follows: less than 10 per cent fines—no qualification; 10 per cent or more but less than 20 per cent fines—'clayey'; 20 to 40 per cent fines—'very clayey'.

The term 'clay' (as written, with single quote marks) is used to describe all material passing $\frac{1}{16}$ mm. Thus it has no mineralogical significance and includes particles falling within the size range of silt. The normal meaning applies to the term clay where it does not appear in single quotation marks.

The ratio of sand to gravel defines the boundaries between sand, pebbly sand, sandy gravel and gravel (at 19:1, 3:1 and 1:1).

Thus it is possible to classify the mineral into one of twelve descriptive categories (see Figure 8). The procedure is as follows:

1 Classify according to ratio of sand to gravel.
2 Describe fines.

For example, a deposit grading 11 per cent gravel, 70 per cent sand and 19 per cent fines is classified as 'clayey' pebbly sand. This short description is included in the borehole log (see Note 12, p. 19).

Many differing proposals exist for the classification of the grain size of sediments (Atterberg, 1905; Udden, 1914; Wentworth, 1922; Wentworth, 1935; Allen, 1936; Twenhofel, 1937; Lane and others, 1947). As Archer (1970a, b) has emphasised, there is a pressing need for a simple metric scale acceptable to both scientific and engineering interests, for which the class limit sizes correspond closely with certain marked changes in the natural properties of mineral particles. For example, there is an important change in the degree of cohesion between particles at about the $\frac{1}{16}$-mm size, which approximates to the generally accepted boundary between silt and sand. These and other requirements are met by a system based on Udden's geometric scale and a simplified form of Wentworth's terminology (Table 9), which is used in this Report.

The fairly wide intervals in the scale are consistent with the general level of accuracy of the qualitative assessments of the resource blocks. Three sizes of sand are recognised, fine ($+\frac{1}{16}$–$\frac{1}{4}$ mm), medium ($+\frac{1}{4}$–1 mm) and coarse (+1–4 mm). The boundary at 16 mm distinguishes a range of finer gravel (+4–16 mm), often characterised by abundance of worn tough pebbles of vein-quartz, from larger pebbles often of notably different materials. The boundary at 64 mm distinguishes pebbles from cobbles. The term 'gravel' is used loosely to denote both pebble-sized and cobble-sized material.

The size distribution of borehole samples is determined by sieve analysis, which is presented by the laboratory as logarithmic cumulative curves (see, for example, British Standard

1377: 1967). In this report the grading is tabulated on the borehole record sheets (Appendix D), the intercepts corresponding with the simple geometric scale $\frac{1}{16}$ mm, $\frac{1}{4}$ mm, 1 mm, 4 mm, 16 mm and so on as required. Original sample grading curves are available for reference at the appropriate office of the Institute.

Each bulk sample is described, subjectively, by a geologist at the borehole site. Being based on visual examination, the description of the grading is inexact, the accuracy depending on the experience of the observer. The descriptions recorded are modified, as necessary, when the laboratory results become available.

The relative proportions of the rock types present in the gravel fraction are indicated by the use of the words 'and' or 'with'. For example, 'flint and quartz' indicates very approximate equal proportions with neither constituent accounting for less than about 25 per cent of the whole; 'flint with quartz' indicates that flint is dominant and quartz, the principal accessory rock type, comprises 5 to 25 per cent of the whole. Where the accessory material accounts for less than 5 per cent of the whole, but is still readily apparent, the phrase 'with some' has been used. Rare constituents are referred to as 'trace'.

The terms used in the field to describe the degree of rounding of particles, which is concerned with the sharpness of the edges and corners of a clastic fragment and not the shape (after Pettijohn, 1957), are as follows.

Angular: showing little or no evidence of wear; sharp edges and corners.

Subangular: showing definite effects of wear. Fragments still have their original form but edges and corners begin to be rounded off.

Subrounded: showing considerable wear. The edges and corners are rounded off to smooth curves. Original grain shape is still distinct.

Rounded: original faces almost completely destroyed, but some comparatively flat surfaces may still remain. All original edges and corners have been smoothed off to rather broad curves. Original shape is still apparent.

Well-rounded: no original faces, edges or corners left. The entire surface consists of broad curves; flat areas are absent. The original shape is suggested by the present form of the grain.

Table 9 Classification of gravel, sand and fines

Size limits	Grain size description	Qualification	Primary classification
	Cobble		
64 mm –			
		Coarse	Gravel
16 mm –	Pebble		
		Fine	
4 mm –			
		Coarse	
1 mm –			
	Sand	Medium	Sand
$\frac{1}{4}$ mm –			
		Fine	
$\frac{1}{16}$ mm –			
	Fines (silt and clay)		Fines

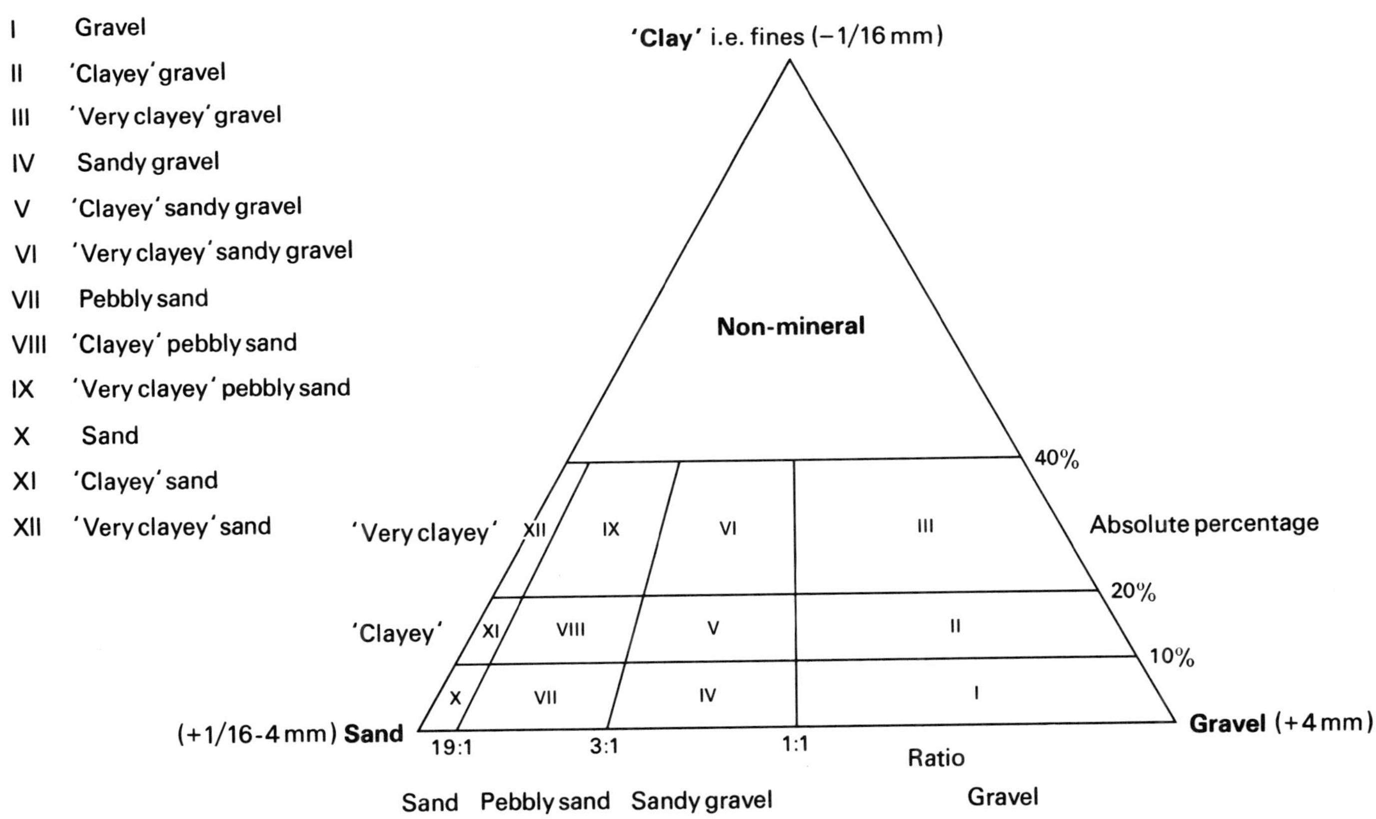

Figure 8 Diagram to show the descriptive categories used in the classification of sand and gravel

APPENDIX D
EXPLANATION OF THE BOREHOLE RECORDS

Annotated Example

TL 51 SE 40[1] **5888 1118**[2] **Frayes**[3] **Block A**

Surface level (+56.7 m) +186 ft[4]
Water struck at +54.5 m and +52.0 m[5]
152 mm percussion[6]
August 1976

Overburden[7] 1.6
Mineral 1.1
Waste 2.0
Mineral 3.0
Bedrock 0.4+[8]

LOG

Geological classification	Lithology	Thickness m	Depth m
	Soil	0.5	0.5
Alluvium[9]	Peat, silty, dark brown	1.1	1.6
	a 'Clayey' sandy gravel[10] Gravel: fine and coarse, angular with well-rounded flint, rounded to subrounded quartzite, quartz and some chalk Sand: medium and coarse with fine, orange-grey	1.1	2.7
Boulder Clay	Clay, olive-grey with chalk, flint and shale pebbles, becoming olive-black from 4.3 m	2.0	4.7
Glacial Sand and Gravel	**b** Sandy gravel Gravel: fine and coarse, angular to subangular and well-rounded flint with some rounded to subrounded quartzite and quartz Sand: medium and fine with coarse, orange-grey	3.0	7.7
London Clay	Clay, stiff, olive-grey, top weathered to brown	0.4+	8.1

GRADING

	Mean for deposit[14] *percentages* Fines	Sand	Gravel	Depth below surface (m)[11]	*percentages*[12] Fines $-\frac{1}{16}$	Sand $+\frac{1}{16}-\frac{1}{4}$	$+\frac{1}{4}-1$	$+1-4$	Gravel $+4-16$	$+16-64$
a	13	51	36	*1.6–2.7	13	6	30	15	25	11
b	1	72	27	*4.7–5.7	1	11	30	8	28	22
				*5.7–6.7	1	32	37	5	11	14
				*6.7–7.7	[3	49	40	3	2	3][13]
				Mean	1	31	36	5	14	13
a+b	4	66	30	Mean	4	24	34	8	17	13

COMPOSITION[15]

	Depth below surface (m)	Percentages by weight in gravel fraction: Angular flint	Well-rounded flint	Quartzite	Quartz	Sandstone	Chalk	Others
a	1.6–2.7	65	8	14	7	0	4	2
b	4.7–7.7	56	35	5	3	0	0	1
a+b	Mean	59	28	7	4	0	1	1

The numbered paragraphs below correspond with the annotations given on the specimen record.

1 Borehole Registration Number
Each Industrial Minerals Assessment Unit (IMAU) borehole is identified by a Registration Number. This consists of two statements.

1 The number of the 1:25 000 sheet on which the borehole lies, for example TL 51.
2 The quarter of the 1:25 000 sheet on which the borehole lies and the number of the borehole in a series for that quarter, for example SE 40.

Thus the full Registration Number is TL 51 SE 40. Usually this is abbreviated to 51 SE 40 in the text.

2 The National Grid reference
All National Grid references in this publication lie within the 100-km square TL unless otherwise stated. Grid references are given to eight figures, accurate to within 10 m for borehole locations. (In the text, six-figure grid references are used for more approximate locations, for example, for farms).

3 Location
The position of the borehole is generally referred to the nearest named locality on the 1:25 000 base map and the resource block in which it lies is stated.

4 Surface level
The surface level at the borehole site is given in metres and feet above Ordnance Datum. All measurements were made in feet; approximate conversions to metres are given in brackets.

5 Groundwater conditions
If groundwater was present, the level at which it was encountered is normally given (in metres above Ordnance Datum).

6 Type of drill and date of drilling
The type of machine, the diameter of the casing used, and the month and year of completion of the borehole are stated.

7 Overburden, Mineral, Waste and Bedrock
Mineral is sand and gravel which, as part of a deposit, falls within the arbitrary definition of potentially workable material (see p. 1). Bedrock is the 'formation', 'country rock' or 'rock head' below which potentially workable sand and gravel will not be found. Waste is any material other than bedrock or mineral. Where waste occurs between the surface and mineral it is classified as overburden.

8 The plus sign (+) indicates that the base of the deposit was not reached during drilling.

9 Geological classification
The geological classification is given wherever possible.

10 Lithological description
When sand and gravel is recorded a general description based on the mean grading characteristics (for details see Appendix C) is followed by more detailed particulars of the gravel and/or sand fractions. Where more than one mineral horizon is recognised, each is designated by a letter, **a**, **b**, etc. The description of other rocks is based on visual examination, in the field.

11 Sampling
A continuous series of bulk samples is taken throughout the thickness of sand and gravel. A new sample is commenced whenever there is an appreciable lithological change within the sand and gravel or at every 1 m of depth. Samples obtained by the bailing technique (that is, from deposits below the water table) are indicated by an asterisk.

12 Grading results
The limits are as follows: gravel, +4 mm, sand, $+\frac{1}{16}$ -4 mm, fines, $-\frac{1}{16}$ mm. Unless otherwise stated all the material passes the 64 mm sieve.

13 If, exceptionally, grading results are not available, an attempt is made to give grading information by comparing the grading and field descriptions of adjacent samples with the samples in question. Such estimates are shown in brackets.

14 Mean grading
The grading of the full thickness of the mineral horizon identified in the log is the mean of the individual sample gradings weighted by the thicknesses represented, if these vary. The classification used is shown in Table 9.

15 Composition
The gravel fraction (+4 mm) is divided into six major rock types in this area. Other rock types occur in minor quantities and are included under 'Others'; these include metamorphic, igneous and ironstone pebbles.

Fully representative sampling of sand and gravel is difficult to achieve particularly where groundwater levels are high. Comparison between boreholes and adjacent exposures suggests that in borehole samples the proportion of sand may be higher and the proportions of fines and coarse gravel (+16 mm) may be lower.

APPENDIX E:

LIST OF BOREHOLES USED IN THE ASSESSMENT OF RESOURCES

Borehole number*	Grid reference†	Page
IMAU BOREHOLES		
51 NW 53	5038 1969	21
54	5024 1838	21
55	5066 1737	22
56	5034 1676	23
57	5026 1553	23
58	5151 1991	24
59	5182 1867	24
60	5180 1529	24
61	5244 1706	25
62	5208 1648	26
63	5346 1814	26
64	5375 1681	26
65	5368 1579	27
66	5466 1990	27
67	5441 1733	28
51 NE 4	5592 1951	28
5	5544 1834	28
6	5687 1978	29
7	5658 1677	29
8	5743 1904	30
9	5742 1684	30
10	5741 1577	31
11	5871 1929	31
12	5831 1833	32
13	5865 1739	32
14	5812 1639	33
15	5820 1502	34
16	5983 1945	35
17	5927 1845	36
18	5971 1751	37
19	5925 1657	37
20	5916 1553	38
51 SW 21	5092 1370	38
22	5098 1234	38
23	5097 1032	39
24	5145 1120	39
25	5258 1497	40
26	5263 1401	41
27	5233 1305	42
28	5232 1217	42
29	5293 1207	43
30	5267 1116	44
31	5235 1031	44
32	5343 1481	45
33	5350 1115	45
34	5385 1052	47
35	5452 1366	48
51 SW 36	5460 1225	48
37	5462 1106	49
38	5471 1021	50
51 SE 30	5551 1150	50
31	5573 1030	51
32	5681 1258	51
33	5628 1099	52
34	5767 1408	52
35	5764 1213	52
36	5794 1142	53
37	5720 1043	54
38	5837 1316	54
39	5868 1219	55
40	5888 1118	56
41	5824 1048	57
42	5956 1301	58
43	5963 1184	58
44	5925 1035	59
61 NW 3	6070 1966	60
4	6022 1852	60
5	6083 1767	61
6	6031 1658	61
7	6052 1536	62
8	6143 1861	62
9	6130 1611	63
10	6298 1974	64
11	6256 1853	65
12	6201 1758	66
13	6282 1697	67
14	6244 1541	67
15	6343 1879	67
16	6338 1783	68
17	6432 1872	69
18	6432 1755	70
19	6408 1623	70
61 NE 18	6539 1963	71
19	6537 1834	72
20	6528 1690	73
21	6502 1563	74
22	6641 1961	75
23	6664 1860	75
24	6675 1715	76
25	6634 1608	77
26	6756 1950	77
27	6737 1865	77
28	6789 1793	78
29	6767 1661	78
61 NE 30	6742 1564	79
31	6853 1975	80
32	6825 1895	81
33	6870 1818	81
34	6859 1642	82
35	6837 1550	83
36	6957 1940	84
37	6969 1792	84
38	6963 1672	85
39	6974 1675	85
61 SW 13	6027 1110	85
14	6160 1477	86
15	6201 1225	86
16	6388 1485	87
17	6385 1126	87
18	6311 1053	87
61 SE 3	6514 1444	88
4	6517 1139	88
5	6502 1018	88
6	6603 1498	89
7	6627 1402	90
8	6655 1306	90
9	6651 1184	90
10	6724 1465	91
11	6754 1368	92
12	6750 1241	92
13	6767 1124	93
14	6716 1022	93
15	6836 1446	94
16	6841 1314	94
17	6818 1037	95
18	6939 1477	96
19	6981 1239	97
20	6962 1141	97
21	6966 1028	98

OTHER BOREHOLES

Other IGS registered boreholes: TL 51 NW 12, 13, 15, 16, 34, 40, 41, 44, 52, 69, 70, 72; TL 51 NE 25; TL 51 SE 46, 48; TL 61 NW 1, 2; TL 61 NE 12, 13; TL 61 SW.20; TL 61 SE 2.

Confidential records: many records made available by the sand and gravel industry are held in confidence.

* By sheet quadrant.

† All fall in 100-km square TL.

APPENDIX F

INDUSTRIAL MINERALS ASSESSMENT UNIT BOREHOLE RECORDS

TL 51 NW 53 5038 1969 Great Hallingbury **Block A**

Surface level (+63.7 m) +209 ft
Water struck at +57.7 m
152 mm percussion
July 1976

Waste 4.2
Bedrock 8.8+

LOG

Geological classification	Lithology	Thickness m	Depth m
	Soil	0.3	0.3
Head	Clay, sandy, orange-brown with flint and quartz pebbles	3.9	4.2
Lower London Tertiaries	Clay, with fine sand laminae, mottled orange-brown and grey	1.5	5.7
	'Clayey' sand: fine with some medium quartz, yellow-orange	6.3	12.0
	'Very clayey' sand: medium with coarse and fine, mottled green and grey	1.0+	13.0

TL 51 NW 54 5024 1838 Victoria Park **Block A**

Surface level (+67.7 m) +222 ft
Water struck at +56.0 m
152 mm percussion
June 1976

Waste 14.8
Bedrock 0.6+

LOG

Geological classification	Lithology	Thickness m	Depth m
	Soil	0.1	0.1
Boulder Clay	Clay, sandy, brown with flints	0.8	0.9
	Clay, mottled orange-grey and light grey with chalk and flint pebbles, becoming grey from 5.2 m with additional pebbles of shale, siltstone, sandstone and quartz below 10.3 m	10.8	11.7
Glacial Sand and Gravel	Gravel Gravel: coarse and fine, angular to subangular flint with rounded to subrounded quartzite, quartz, some well-rounded flint and sandstone Sand: medium with coarse and fine, quartz with some flint, orange-grey	3.1	14.8
London Clay	Clay, micaceous, blue-black, top 0.2 m weathered to brown	0.6+	15.4

GRADING

Mean for deposit *percentages*			Depth below surface (m)	*percentages*					
Fines	Sand	Gravel		Fines	Sand			Gravel	
				$-\frac{1}{16}$	$+\frac{1}{16}-\frac{1}{4}$	$+\frac{1}{4}-1$	+1–4	+4–16	+16–64
1	46	53	*11.7–12.7	2	3	56	10	17	12
			*12.7–13.7	1	2	32	9	26	30
			*13.7–14.8	1	2	19	6	25	47
			Mean	1	3	35	8	23	30

COMPOSITION

Depth below surface (m)	Percentages by weight in gravel fraction						
	Angular flint	Well-rounded flint	Quartzite	Quartz	Sandstone	Chalk	Others
11.7–14.8	52	5	20	15	3	0	5

TL 51 NW 55 5066 1737 Wrights' Green Farm **Block A**

Surface level (+68.9 m) +226 ft
Water struck at +52.8 m
152 mm percussion
May 1977

Waste 19.0
Bedrock 1.1+

LOG

Geological classification	Lithology	Thickness m	Depth m
	Soil	0.3	0.3
Boulder Clay	Clay, brown with chalk and flint pebbles, becoming mottled orange-grey and light grey from 1.0 m, passing to grey from 5.7 m with a seam of chalky sand from 7.9 to 8.3 m	15.8	16.1
Glacial Sand and Gravel	Gravel, with pebbly clay from 17.0 to 17.2 m Gravel: fine and coarse, angular to subangular flint with chalk and quartzite Sand, coarse and medium with some fine, quartz with chalk and flint	2.9	19.0
London Clay	Silt, clayey and sandy, olive-grey	1.1+	20.1

GRADING

Mean for deposit *percentages*			Depth below surface (m)	*percentages*					
Fines	Sand	Gravel		Fines	Sand			Gravel	
				$-\frac{1}{16}$	$+\frac{1}{16}-\frac{1}{4}$	$+\frac{1}{4}-1$	+1–4	+4–16	+16–64
4	35	61	*16.1–17.0	5	1	3	10	46	35
			17.0–17.2	Pebbly clay					
			*17.2–19.0	4	1	15	29	32	19
			Mean	4	1	11	23	37	24

TL 51 NW 56 5034 1676 Grinsted Lane **Block A**

Surface level (+60.7 m) +199 ft
Water struck at +51.7 m
152 mm percussion
May 1977

Waste 10.0
Bedrock 2.2+

LOG

Geological classification	Lithology	Thickness m	Depth m
	Soil and subsoil	0.6	0.6
Head	Clay, brown with flints, becoming sandy with many pebbles towards base	0.8	1.4
Boulder Clay	Clay, brown with chalk and flint pebbles, becoming grey with sandy seams	7.6	9.0
Glacial Sand and Gravel	Gravel Gravel: fine and coarse, angular to subangular with some well-rounded flint and chalk Sand: coarse with medium and some fine, quartz with flint and some chalk	1.0	10.0
London Clay	Clay, silty and sandy, olive-grey	2.2+	12.2

GRADING

Mean for deposit *percentages*			Depth below surface (m)	*percentages*					
Fines	Sand	Gravel		Fines	Sand			Gravel	
				$-\frac{1}{16}$	$+\frac{1}{16}-\frac{1}{4}$	$+\frac{1}{4}-1$	+1–4	+4–16	+16–64
6	30	64	*9.0–10.0	6	1	7	22	44	20

TL 51 NW 57 5026 1553 Little Hyde Hall **Block A**

Surface level (+69.8 m) +229 ft
Water struck at +68.4 m
152 mm percussion
June 1976

Waste 18.6+

LOG

Geological classification	Lithology	Thickness m	Depth m
	Soil	0.3	0.3
Boulder Clay	Clay, silty, mottled orange-grey and light grey with chalk and flint pebbles	1.1	1.4
	Silt, sandy, orange-grey with many chalk and flint pebbles	1.6	3.0
	Clay, silty, light grey with chalk pebbles, becoming olive-grey with additional pebbles of shale from 4.5 m. Passing down into clay, stiff, dark grey with fine pebbles of chalk, shale and flint below 14.5 m	15.6+	18.6

TL 51 NW 58 5151 1991 Long Plantation **Block A**

Surface level (+89.6 m) +294 ft Waste 18.6+
Water not struck
152 mm percussion
July 1976

LOG

Geological classification	Lithology	Thickness m	Depth m
	Soil	0.1	0.1
Boulder Clay	Clay, brown with chalk and flint pebbles, becoming mottled orange-grey and light grey with sandy seams from 1.8 m	6.8	6.9
	Sand, silty, orange-grey with many chalk and flint pebbles	1.7	8.6
	Clay, olive-grey with chalk, flint and shale pebbles. Passing down into clay, stiff, dark grey with fine pebbles of chalk, sandstone, ironstone, siltstone, shale and flint below 14.0 m	10.0+	18.6

TL 51 NW 59 5182 1867 Ladywell **Block A**

Surface level (+88.4 m) +290 ft Waste 19.2+
Water not struck
152 mm percussion
August 1977

LOG

Geological classification	Lithology	Thickness m	Depth m
	Soil	0.4	0.4
Boulder Clay	Clay, brown with flints	0.6	1.0
	Clay, brown with chalk and flint pebbles, becoming olive-grey towards base	6.3	7.3
	Sand, silty, olive-grey with some fine chalk pebbles	2.3	9.6
	Clay, light grey with chalk and flint pebbles, becoming olive-grey from 10.5 m with additional pebbles of shale below 18.5 m	9.6+	19.2

TL 51 NW 60 5180 1529 Hatfield Heath **Block A**

Surface level (+80.8 m) +265 ft Waste 5.4+
Water not struck
152 mm percussion
January 1977

LOG

Geological classification	Lithology	Thickness m	Depth m
	Made ground	0.5	0.5
Head Gravel	Clay, very sandy, mottled orange-brown and light grey, with flints, becoming more sandy below 1.3 m	1.5	2.0
Boulder Clay	Clay, mottled orange-grey and light grey with chalk, flint, sandstone and shale pebbles, becoming brown towards base	3.4+	5.4

TL 51 NW 61 5244 1706 Ryes **Block A**

Surface level (+78.3 m) +257 ft
Water struck at +56.3 m
152 mm percussion
November 1976

Overburden 15.6
Mineral 7.5
Bedrock 1.1+

LOG

Geological classification	Lithology	Thickness m	Depth m
	Soil	0.2	0.2
Boulder Clay	Clay, mottled grey and brown	0.7	0.9
	Sand, fine, orange-yellow	0.5	1.4
	Clay, mottled olive-brown and light grey with chalk and flint pebbles, chalk gravel from 1.7 to 2.3 m, becoming olive-grey with additional pebbles of siltstone and quartzite below 5.5 m	14.2	15.6
Glacial Sand and Gravel	Gravel Gravel: fine and coarse, angular to subangular with well-rounded flint, rounded to subrounded quartzite and quartz Sand: medium with coarse and fine, quartz with flint	7.5	23.1
London Clay	Clay, silty grey, top 0.2 m weathered to orange-brown	1.1+	24.2

GRADING

Mean for deposit *percentages*			Depth below surface (m)	*percentages*					
Fines	Sand	Gravel		Fines	Sand			Gravel	
				$-\frac{1}{16}$	$+\frac{1}{16}-\frac{1}{4}$	$+\frac{1}{4}-1$	+1–4	+4–16	+16–64
6	45	49	*15.6–16.6	18	11	36	7	19	9
			*16.6–17.6	7	8	47	7	17	14
			*17.6–18.6	5	23	34	6	18	14
			*18.6–19.6	8	13	28	12	28	11
			*19.6–20.6	2	2	11	14	47	24
			*20.6–21.6	3	3	13	10	40	31
			*21.6–22.6	4	3	16	10	42	25
			*22.6–23.1	6	7	23	14	24	26
			Mean	6	9	26	10	30	19

COMPOSITION

Depth below surface (m)	Percentages by weight in gravel fraction						
	Angular flint	Well-rounded flint	Quartzite	Quartz	Sandstone	Chalk	Others
15.6–23.1	53	13	23	8	1	0	2

TL 51 NW 62 5208 1648 Little Hallingbury Park **Block A**

Surface level (+72.2 m) +237 ft
Water not struck
152 mm percussion
May 1977

Waste 18.3+

LOG

Geological classification	Lithology	Thickness m	Depth m
	Soil and subsoil	0.5	0.5
Boulder Clay	Clay, sandy, brown with flints	0.2	0.7
	Clay, mottled orange-brown and light grey with chalk and flint pebbles, becoming grey with additional pebbles of shale from 2.7 m. Passing down into clay, stiff, dark grey, with fine pebbles of chalk, flint, sandstone and siltstone below 11.5 m	17.6+	18.3

TL 51 NW 63 5346 1814 Tom's by the Wood **Block A**

Surface level (+88.7 m) +291 ft
Water not struck
152 mm percussion
November 1976

Waste 18.4+

LOG

Geological classification	Lithology	Thickness m	Depth m
	Soil	0.2	0.2
Boulder Clay	Silt, clayey, brown with flints	0.5	0.7
	Clay, mottled brown and light grey with chalk and flint pebbles, silty sand from 4.5 to 5.0 m, becoming olive-grey from 5.0 m. Passing down into clay, dark grey, with mostly fine pebbles of chalk, shale and flint below 13.0 m	17.7+	18.4

TL 51 NW 64 5375 1681 Bleak House **Block A**

Surface level (+77.1 m) +253 ft
Water not struck
152 mm percussion
November 1976

Waste 18.4+

LOG

Geological classification	Lithology	Thickness m	Depth m
	Soil	0.3	0.3
Boulder Clay	Clay, silty, mottled brown and grey with flints, gastropod shells and plant debris	0.8	1.1
	Clay, silty, brown with chalk and flint pebbles, becoming clay, olive-grey with additional pebbles of quartz and siltstone from 4.0 m. Passing down into clay, dark grey with mostly fine pebbles of chalk, flint and siltstone below 14.1 m	17.3+	18.4

TL 51 NW 65 5368 1579 Old Street Hill **Block A**

Surface level (+62.2 m) +204 ft
Water struck at +44.1 m
152 mm percussion
October 1976

Waste 22.2
Bedrock 0.4+

LOG

Geological classification	Lithology	Thickness m	Depth m
	Soil	0.3	0.3
Alluvium	Clay, silty, brown with gastropod shells	1.7	2.0
Boulder Clay	Clay, mottled grey and brown with chalk and flint pebbles, becoming olive-grey with additional pebbles of siltstone and quartzite from 2.9 m, sand from 5.8 to 6.3. Passing down into clay, dark grey with mostly fine pebbles of chalk, flint, shale, siltstone and sandstone below 15.0 m	16.1	18.1
Glacial Sand and Gravel	Gravel Gravel: fine and coarse, subangular to subrounded flint with quartz Sand: medium and coarse with some fine, quartz with flint	4.1	22.2
London Clay	Silt, clayey, micaceous, grey	0.4+	22.6

GRADING

Mean for deposit *percentages* Fines	Sand	Gravel	Depth below surface (m)	*percentages* Fines $-\frac{1}{16}$	Sand $+\frac{1}{16}-\frac{1}{4}$	$+\frac{1}{4}-1$	$+1-4$	Gravel $+4-16$	$+16-64$
3	38	59	*18.1–19.1	3	2	21	13	41	20
			*19.1–22.2	[3	3	23	15	38	18]
			Mean	3	2	22	14	40	19

TL 51 NW 66 5466 1990 Gravelpit Coppice **Block A**

Surface level (+87.2 m) +286 ft
Water not struck
152 mm percussion
November 1976

Waste 18.4+

LOG

Geological classification	Lithology	Thickness m	Depth m
	Soil	0.1	0.1
Boulder Clay	Clay, silty light brown with plant debris	0.5	0.6
	Clay, mottled grey and brown with chalk and flint pebbles, becoming olive-grey from 4.6 m with additional pebbles of siltstone, quartz, sandstone and shale from 9.0 m, passing to olive-black with fewer chalk pebbles towards base	17.8+	18.4

TL 51 NW 67 **5441 1733** **The Marsh** **Block A**

Surface level (+65.2 m) +214 ft
Water struck at +56.5 m
152 mm percussion
October 1976

Waste 9.1
Bedrock 1.9+

LOG

Geological classification	Lithology	Thickness m	Depth m
	Soil and subsoil	1.5	1.5
Alluvium	Clay, silty, grey, becoming peaty with plant debris and fragile shells	6.6	8.1
Boulder Clay	Clay, dark grey with chalk, siltstone and flint pebbles	0.6	8.7
Glacial Sand and Gravel	Gravel Gravel: fine with coarse, angular flint with quartz Sand: coarse, with some medium and fine, quartz with flint	0.4	9.1
London Clay	Silt, sandy, micaceous, olive-grey	1.9+	11.0

TL 51 NE 4 **5592 1951** **Hart Farm** **Block B**

Surface level (+100.6 m) +330 ft
Water not struck
152 mm percussion
January 1977

Waste 19.4+

LOG

Geological classification	Lithology	Thickness m	Depth m
	Made ground	0.4	0.4
Boulder Clay	Clay, mottled orange-brown and light grey with chalk pebbles, becoming olive-grey from 3.2 m. Passing down into clay, stiff with fine pebbles of chalk, shale, flint and fossil fragment below 18.7 m	19.0+	19.4

TL 51 NE 5 **5544 1834** **Brumsted Common** **Block A**

Surface level (+82.0 m) +269 ft
Water not struck
152 mm percussion
August 1976

Waste 16.5
Bedrock 0.2+

LOG

Geological classification	Lithology	Thickness m	Depth m
	Soil	0.2	0.2
Boulder Clay	Clay, mottled light grey and brown with chalk and siltstone pebbles, becoming olive-grey with additional pebbles of flint, sandstone, quartz and shale from 4.6 m. Passing down into clay, stiff, olive-black with mostly fine pebbles of chalk, flint, shale, ironstone, siltstone and fossil fragment below 5.8 m	10.0	10.2
	Clay, very sandy, stiff, olive-brown with many pebbles of flint and quartzite	6.3	16.5
London Clay	Clay, micaceous, olive-grey	0.2+	16.7

TL 51 NE 6 5687 1978 Puttock's End **Block B**

Surface level (+93.6 m) +307 ft
Water struck at +77.2 m
152 mm percussion
December 1976

Overburden 13.8
Mineral 8.7
Bedrock 0.6+

LOG

Geological classification	Lithology	Thickness m	Depth m
	Made ground	0.2	0.2
Boulder Clay	Clay, mottled brown and grey with chalk and flint pebbles, becoming olive-grey with additional pebbles of shale and sandstone from 7.0 m. Passing down into clay, olive-black with fine pebbles of chalk, flint and shale below 10.4 m	12.5	12.7
	Clay, sandy, brown with quartz and flint pebbles	1.1	13.8
Glacial Sand and Gravel	Gravel	8.7	22.5
	Gravel: fine and coarse, angular to subangular with well-rounded flint and rounded quartzite and quartz		
	Sand: medium and coarse with fine, quartz with flint		
London Clay	Clay, olive-grey, top weathered to brown	0.6+	23.1

GRADING

Mean for deposit *percentages* Fines	Sand	Gravel	Depth below surface (m)	*percentages* Fines $-\frac{1}{16}$	Sand $+\frac{1}{16}-\frac{1}{4}$	$+\frac{1}{4}-1$	$+1-4$	Gravel $+4-16$	$+16-64$
6	46	48	*13.8–14.8	8	5	17	17	41	12
			*14.8–15.8	5	4	21	8	32	30
			*15.8–16.8	5	5	23	14	32	21
			*16.8–17.8	8	6	33	17	23	13
			*17.8–18.8	9	12	41	15	13	10
			*18.8–19.8	4	10	17	29	16	24
			*19.8–20.8	5	5	31	10	17	32
			*20.8–21.9	2	2	17	10	26	43
			*21.9–22.5	4	3	29	17	26	21
			Mean	6	6	25	15	25	23

COMPOSITION

Depth below surface (m)	Percentages by weight in gravel fraction: Angular flint	Well-rounded flint	Quartzite	Quartz	Sandstone	Chalk	Others
13.8–22.5	41	17	19	19	1	0	3

TL 51 NE 7 5658 1677 Loftus Lodge **Block B**

Surface level (+75.0 m) +246 ft
Water not struck
152 mm percussion
September 1976

Waste 18.6+

LOG

Geological classification	Lithology	Thickness m	Depth m
	Soil	0.3	0.3
Boulder Clay	Clay, mottled orange-brown and light grey with chalk and flint pebbles, becoming grey from 4.2 m. Passing down into clay, stiff, olive-grey with fine pebbles of chalk, shale, siltstone, flint and fossil fragment below 8.8 m	18.3+	18.6

TL 51 NE 8 5743 1904 Dealtree Farm **Block B**

Surface level (+87.8 m) +288 ft
Water struck at +75.7 m
152 mm percussion
December 1976

Waste 14.6
Bedrock 1.2+

LOG

Geological classification	Lithology	Thickness m	Depth m
	Soil and subsoil	0.6	0.6
Boulder Clay	Clay, mottled brown and orange-grey with chalk and flint pebbles, becoming mottled olive-grey and brown from 4.0 m. Passing down into clay, dark grey with fine pebbles of chalk, flint, sandstone, siltstone, shale and quartzite below 6.0 m	11.5	12.1
Glacial Sand and Gravel	Gravel Gravel: coarse and fine, angular to subangular flints with rounded quartzite and quartz Sand: medium and coarse with fine, quartz and flint	2.5	14.6
London Clay	Clay, silty, dark grey, top weathered to brown	1.2+	15.8

GRADING

Mean for deposit *percentages*			Depth below surface (m)	*percentages*					
Fines	Sand	Gravel		Fines	Sand			Gravel	
				$-\frac{1}{16}$	$+\frac{1}{16}-\frac{1}{4}$	$+\frac{1}{4}-1$	+1–4	+4–16	+16–64
5	44	51	*12.1–13.1	5	4	16	12	29	34
			*13.1–14.6	5	3	36	13	23	20
			Mean	5	3	28	13	25	26

TL 51 NE 9 5742 1684 Broomshawbury **Block B**

Surface level (+75.3 m) +247 ft

Water not struck
152 mm percussion
December 1976

Waste 11.8
Bedrock 5.2+

LOG

Geological classification	Lithology	Thickness m	Depth m
	Made ground	0.3	0.3
Boulder Clay	Clay, mottled light grey to orange-grey with chalk and flint pebbles, becoming grey with sandy seams from 3.9 m. Passing down into clay, stiff, olive-black with flint, siltstone, chalk and fossil fragment pebbles below 7.2 m	9.6	9.9
	Clay, sandy, green-grey with flints	1.9	11.8
London Clay	Clay, micaceous, light brown, becoming olive-grey towards base	5.2+	17.0

TL 51 NE 10 5741 1577 Poplars Wood **Block B**

Surface level (+73.2 m) +240 ft
Water not struck
152 mm percussion
October 1976

Waste 18.7+

LOG

Geological classification	Lithology	Thickness m	Depth m
	Soil	0.5	0.5
Boulder Clay	Clay, brown with flint and chalk pebbles	2.1	2.6
	Clay, mottled orange-grey and light grey with chalk and flint pebbles, becoming olive-grey from 4.8 m. Passing down into clay, stiff, olive-black with chalk, flint and shale pebbles below 13.4 m	16.1+	18.7

TL 51 NE 11 5871 1929 Ashfield **Block B**

Surface level (+74.4 m) +244 ft
Water not struck
152 mm percussion
December 1976

Waste 11.9
Bedrock 0.6+

LOG

Geological classification	Lithology	Thickness m	Depth m
	Soil	0.2	0.2
Alluvium	Clay, silty, mottled brown and grey with pebbles and plant debris	1.2	1.4
	Gravel Gravel: fine with coarse, angular to subangular flint with quartz and some chalk Sand: medium	0.2	1.6
Boulder Clay	Clay, olive-grey with fine pebbles of chalk, flint, shale, quartz, quartzite and sandstone, becoming olive-black below 8.4 m	8.2	9.8
	Clay, sandy, olive-brown with many pebbles of flint and quartz	2.1	11.9
London Clay	Clay, micaceous, olive-grey, top weathered to olive- brown	0.6+	12.5

TL 51 NE 12 5831 1833 Black Hall **Block B**

Surface level (+84.1 m) +276 ft
Water struck at +75.0 m
152 mm percussion
October 1976

Waste 11.1
Bedrock 1.0+

LOG

Geological classification	Lithology	Thickness m	Depth m
	Soil	0.3	0.3
Boulder Clay	Clay, brown with chalk and flint pebbles, mottled to light grey from 1.3 m, becoming grey from 4.8 m. Passing down into clay, stiff, dark grey with fine chalk, flint, siltstone, sandstone and shale pebbles below 5.3 m	8.7	9.0
Glacial Sand and Gravel	Sandy gravel, with 'clayey' sand from 9.0 to 9.1 m Gravel: fine and coarse, subangular with well-rounded flint, rounded quartzite, some quartz and sandstone Sand: medium and coarse with fine quartz	2.1	11.1
London Clay	Clay, stiff, olive-grey, top 0.4 m weathered to orange-brown	1.0+	12.1

GRADING

Mean for deposit *percentages*			Depth below surface (m)	*percentages*						
Fines	Sand	Gravel		Fines	Sand			Gravel		
				$-\frac{1}{16}$	$+\frac{1}{16}-\frac{1}{4}$	$+\frac{1}{4}-1$	+1–4	+4–16	+16–64	+64
4	53	43	* 9.0–10.0	7	4	13	17	37	22	0
			*10.0–11.1	2	15	43	11	13	7	9
			Mean	4	10	29	14	24	14	5

COMPOSITION

Depth below surface (m)	Percentages by weight in gravel fraction						
	Angular flint	Well-rounded flint	Quartzite	Quartz	Sandstone	Chalk	Others
9.0–11.1	51	16	25	4	2	0	2

TL 51 NE 13 5865 1739 Marsh Farm **Block B**

Surface level (+80.5 m) +264 ft
Water not struck
152 mm percussion
September 1976

Waste 18.6+

LOG

Geological classification	Lithology	Thickness m	Depth m
	Soil	0.3	0.3
Boulder Clay	Clay, mottled brown and light grey with chalk pebbles	1.2	1.5
	Silt, interlaminated with fine sand, orange-grey with chalk pebbles towards base	5.3	6.8
	Clay, olive-grey, with chalk pebbles, becoming dark grey below 16.5 m	11.8+	18.6

TL 51 NE 14 5812 1639 Newhall **Block B**

Surface level (+66.1 m) +217 ft
Water struck at +50.2 m
152 mm percussion
September 1976

Overburden 1.3
Mineral 1.0
Waste 11.2
Mineral 8.3
Bedrock 1.2+

LOG

Geological classification	Lithology	Thickness m	Depth m
	Soil	0.3	0.3
Alluvium	Clay, silty, mottled brown and grey with plant debris, shell fragments and pebbles of chalk and flint	1.0	1.3
	a 'Clayey' gravel, with clay seams Gravel: fine and coarse, angular with well-rounded flint, subrounded quartzite, quartz, some rounded chalk and sandstone Sand: medium and coarse with fine	1.0	2.3
Boulder Clay	Clay, olive-grey with chalk, flint and siltstone pebbles	3.1	5.4
	Silt, interlaminated with fine sand	1.1	6.5
	Clay, dark grey with fine chalk and flint pebbles	7.0	13.5
Glacial Sand and Gravel	**b** Pebbly sand, with laminated silt and clay from 14.4 to 14.7 m Gravel: fine and coarse, angular to subangular with well-rounded flint, rounded to subrounded quartzite and quartz Sand: medium and fine with coarse quartz	8.3	21.8
London Clay	Clay, silty, stiff, dark grey, top 0.2 m weathered to brown	1.2+	23.0

GRADING

	Mean for deposit *percentages*			Depth below surface (m)	*percentages*					
	Fines	Sand	Gravel		Fines	Sand			Gravel	
					$-\frac{1}{16}$	$+\frac{1}{16}-\frac{1}{4}$	$+\frac{1}{4}-1$	+1–4	+4–16	+16–64
a	15	31	54	1.3–2.3	15	5	15	11	30	24
b	10	68	22	13.5–14.4	7	40	26	4	10	13
				14.4–14.7	Laminated silt and clay					
				14.7–15.7	8	10	18	14	38	12
				*15.7–16.7	8	35	21	6	15	15
				*16.7–17.7	5	11	64	9	9	2
				*17.7–18.7	5	10	78	4	3	0
				*18.7–19.7	34	11	48	3	2	2
				*19.7–20.7	6	32	39	5	8	10
				*20.7–21.8	6	27	24	6	13	24
				Mean	10	21	41	6	12	10
a+b	10	64	26	Mean	10	19	38	7	14	12

COMPOSITION

	Depth below surface (m)	Percentages by weight in gravel fraction						
		Angular flint	Well-rounded flint	Quartzite	Quartz	Sandstone	Chalk	Others
a	1.3–2.3	55	21	10	7	2	2	3
b	13.5–21.8	57	15	19	6	1	0	2
a+b	Mean	57	16	18	6	1	0	2

TL 51 NE 15   5820 1502   Brooks Farm   **Block B**

Surface (+63.7 m) +209 ft
Water struck at +47.2 m
152 mm percussion
October 1976

Waste 20.0
Bedrock 0.8+

LOG

Geological classification	Lithology	Thickness m	Depth m
	Soil	0.3	0.3
Alluvium	Clay, silty and sandy, brown with flints	1.0	1.3
	'Very clayey' pebbly sand Gravel: fine and coarse, angular flint with quartz and some chalk Sand: medium and coarse with fine	0.8	2.1
Boulder Clay	Clay, mottled orange-grey and light grey with chalk, quartzite, flint and sandstone pebbles, becoming olive-grey from 4.0 m. Passing down into clay, olive-black with fine chalk, flint, shale and fossil fragment pebbles below 12.0 m	14.4	16.5
Glacial Sand and Gravel	Gravel Gravel: fine and coarse, angular to subangular flint with rounded to subrounded quartzite and quartz Sand: medium and coarse with fine, brown-grey	3.5	20.0
London Clay	Clay, silty, olive-grey, top weathered to olive-brown	0.8+	20.8

GRADING

Mean for deposit *percentages*			Depth below surface (m)	*percentages*					
Fines	Sand	Gravel		Fines	Sand			Gravel	
				$-\frac{1}{16}$	$+\frac{1}{16}-\frac{1}{4}$	$+\frac{1}{4}-1$	+1–4	+4–16	+16–64
4	43	53	*16.5–17.5	6	19	34	14	18	9
			*17.5–18.5	3	7	16	9	34	31
			*18.5–20.0	4	6	13	15	35	27
			Mean	4	10	20	13	30	23

TL 51 NE 16 **5983 1945** **The Elms** **Block B**

Surface level (+84.7 m) +278 ft
Water struck at +71.9 m
152 mm percussion
December 1976

Overburden 11.8
Mineral 7.2
Bedrock 0.4+

LOG

Geological classification	Lithology	Thickness m	Depth m
	Made ground	0.7	0.7
Boulder Clay	Clay, light brown with fine pebbles of chalk and flint, becoming grey from 6.0 m. Passing down into clay, dark grey, with fine pebbles of chalk, flint, siltstone, shale, quartzite and quartz below 8.0 m	11.1	11.8
Glacial Sand and Gravel	'Clayey' sandy gravel, with silt seams above 14.0 m Gravel: fine and coarse, angular to subangular with well-rounded flint, rounded to subrounded quartzite, quartz and sandstone Sand: medium with fine and coarse	7.2	19.0
London Clay	Clay, olive-grey, top weathered to olive-brown	0.4+	19.4

GRADING

Mean for deposit *percentages* Fines	Sand	Gravel	Depth below surface (m)	*percentages* Fines $-\frac{1}{16}$	Sand $+\frac{1}{16}-\frac{1}{4}$	$+\frac{1}{4}-1$	$+1-4$	Gravel $+4-16$	$+16-64$
14	54	32	11.8–12.8	28	23	44	1	1	3
			*12.8–14.0	20	12	39	5	12	12
			*14.0–15.0	7	3	29	9	22	30
			*15.0–16.0	27	7	32	6	19	9
			*16.0–17.0	8	3	59	10	15	5
			*17.0–18.0	3	3	26	12	26	30
			*18.0–19.0	4	3	38	12	19	24
			Mean	14	8	38	8	16	16

COMPOSITION

Depth below surface (m)	Percentages by weight in gravel fraction: Angular flint	Well-rounded flint	Quartzite	Quartz	Sandstone	Chalk	Others
11.8–19.0	44	16	21	14	2	0	3

TL 51 NE 17 **5927 1845** **Bury Farm** **Block B**

Surface level (+70.7 m) +232 ft
Water struck at +68.7 m
152 mm percussion
October 1976

Overburden 2.9
Mineral 2.5
Bedrock 1.0+

LOG

Geological classification	Lithology	Thickness m	Depth m
	Soil	0.2	0.2
Alluvium	Peat, clayey, dark brown. Passing down into sand, silty, green-grey with shell fragments below 2.5 m	2.7	2.9
	'Clayey' gravel, with clay seams from 2.9 to 4.3 m Gravel: coarse and fine, angular with well-rounded flint, subrounded quartzite, some quartz and sandstone Sand: fine and medium with coarse, grey	2.5	5.4
London Clay	Clay, silty, stiff, olive-brown	1.0+	6.4

GRADING

Mean for deposit *percentages*			Depth below surface (m)	*percentages*					
Fines	Sand	Gravel		Fines	Sand			Gravel	
				$-\frac{1}{16}$	$+\frac{1}{16}-\frac{1}{4}$	$+\frac{1}{4}-1$	+1–4	+4–16	+16–64
20	31	49	*2.9–4.3	30	26	16	4	8	16
			*4.3–5.4	5	1	4	9	42	39
			Mean	20	15	10	6	23	26

COMPOSITION

Depth below surface (m)	Percentages by weight in gravel fraction						
	Angular flint	Well-rounded flint	Quartzite	Quartz	Sandstone	Chalk	Others
2.9–5.4	62	11	17	4	2	0	4

TL 51 NE 18 5971 1751 High Rodingbury Wood **Block B**

Surface level (+75.3 m) +247 ft
Water not struck
152 mm percussion
September 1976

Overburden 4.3
Mineral 2.8
Bedrock 1.0+

LOG

Geological classification	Lithology	Thickness m	Depth m
	Soil	0.3	0.3
Boulder Clay	Clay, mottled brown and grey with chalk and flint pebbles, becoming dark grey below 2.5 m	4.0	4.3
Glacial Sand and Gravel	Gravel, with clay seams Gravel: coarse and fine, rounded quartzite, angular with well-rounded flint, rounded quartz and some sandstone Sand: medium with coarse and some fine quartz	2.8	7.1
London Clay	Clay, dark grey, top 0.3 m weathered to brown	1.0+	8.1

GRADING

Mean for deposit *percentages*			Depth below surface (m)	*percentages*						
Fines	Sand	Gravel		Fines	Sand			Gravel		
				$-\frac{1}{16}$	$+\frac{1}{16}-\frac{1}{4}$	$+\frac{1}{4}-1$	+1–4	+4–16	+16–64	+64
2	41	57	4.3–6.2	3	2	36	8	23	28	0
			6.2–7.1	1	1	16	10	31	36	5
			Mean	2	2	30	9	25	30	2

COMPOSITION

Depth below surface (m)	Percentages by weight in gravel fraction						
	Angular flint	Well-rounded flint	Quartzite	Quartz	Sandstone	Chalk	Others
4.3–7.1	30	22	37	7	2	0	2

TL 51 NE 19 5925 1657 High Rodingbury **Block B**

Surface (+85.0 m) +279 ft
Water not struck
152 mm percussion
October 1976

Waste 17.3
Bedrock 0.3+

LOG

Geological classification	Lithology	Thickness m	Depth m
	Soil	0.3	0.3
Boulder Clay	Clay, brown with chalk and flint pebbles, becoming grey with additional pebbles of siltstone, from 7.7 m. Passing down into clay, stiff, dark grey with fine pebbles of chalk, siltstone, flint and shale below 14.0 m	17.0	17.3
London Clay	Clay, olive-grey	0.3+	17.6

TL 51 NE 20 **5916 1553** **Gunner's Farm** **Block B**

Surface level (+79.6 m) +261 ft
Water not struck
152 mm percussion
October 1976

Waste 18.6+

LOG

Geological classification	Lithology	Thickness m	Depth m
	Soil	0.4	0.4
Boulder Clay	Clay, brown with chalk and flint pebbles, becoming grey below 10.4 m	18.2+	18.6

TL 51 SW 21 **5092 1370** **Sheering** **Block A**

Surface level (+75.6 m) +248 ft
Water not struck
152 mm percussion
June 1976

Waste 18.6+

LOG

Geological classification	Lithology	Thickness m	Depth m
	Soil	0.1	0.1
Boulder Clay	Clay, orange-brown with chalk and flint pebbles, becoming orange-grey from 3.0 m and olive-grey from 9.5 m with additional pebbles of shale and siltstone below 14.0 m	18.5+	18.6

TL 51 SW 22 **5098 1234** **Chestnut** **Block A**

Surface level (+60.7 m) +199 ft
Water not struck
152 mm percussion
June 1976

Waste 13.4
Bedrock 0.6+

LOG

Geological classification	Lithology	Thickness m	Depth m
	Soil	0.2	0.2
Head	Silt, clayey, mottled olive-grey and brown with fine chalk pebbles towards base	1.6	1.8
Boulder Clay	Clay, mottled brown and olive-grey with chalk pebbles, becoming olive-grey with additional fine pebbles of quartz, sandstone and flint. Passing down into clay, olive-black with fine pebbles of chalk, flint, shale and quartz below 10.0 m	11.2	13.0
	Clay, very sandy, olive-grey with many flint and quartzite pebbles	0.4	13.4
London Clay	Clay, micaceous, olive-grey	0.6+	14.0

TL 51 SW 23 5097 1032 Loyter's Green **Block A**

Surface level (+87.5 m) +287 ft
Water struck at +72.4 m
152 mm percussion

Waste 15.6
Bedrock 1.4+

LOG

Geological classification	Lithology	Thickness m	Depth m
	Soil	0.1	0.1
Boulder Clay	Clay, mottled orange-grey and light grey with chalk and flint pebbles, becoming olive-grey from 4.0 m. Passing down into clay, olive-black with mostly fine pebbles of chalk, shale, sandstone, fossil fragment, flint and quartz below 14.2 m	15.0	15.1
Glacial Sand and Gravel	Sandy gravel Gravel: coarse and fine, angular to subangular with well-rounded flint and rounded to subrounded quartz Sand: coarse and medium with fine quartz, light grey	0.5	15.6
London Clay	Clay, olive-grey, top 1.1 m weathered to brown	1.4+	17.0

TL 51 SW 24 5145 1120 Matching Tye **Block A**

Surface level (+86.9 m) +285 ft
Water not struck
152 mm percussion
June 1976

Waste 18.6+

LOG

Geological classification	Lithology	Thickness m	Depth m
	Soil	0.2	0.2
Boulder Clay	Clay, dark brown with chalk and flint pebbles, mottled orange-grey and light grey from 0.8 m, becoming olive-grey from 7.0 m with additional pebbles of shale towards base	18.4+	18.6

TL 51 SW 25 **5258 1497** **Hatfield Heath** **Block A**

Surface level (+77.7 m) +255 ft
Water struck at +74.7 m
152 mm percussion
January 1977

Overburden 0.5
Mineral 6.0
Waste 17.9+

LOG

Geological classification	Lithology	Thickness m	Depth m
	Soil	0.1	0.1
Head Gravels	Clay, very sandy, mottled light grey and orange-brown with flints	0.4	0.5
	'Clayey' gravel Gravel: coarse and fine, angular to subrounded flint with rounded chalk, some rounded to subrounded quartzite, well-rounded flint, rounded sandstone, angular ironstone and rounded quartz Sand: coarse and medium with fine, orange-brown.	6.0	6.5
Boulder Clay	Clay, brown with chalk pebbles, becoming grey from 8.0 m, 'clayey' chalk gravel from 23.6 to 23.8 m. Passing down into clay, stiff, dark grey with chalk, shale, fossil fragment with siltstone pebbles below 23.8 m	17.9+	24.4

GRADING

Mean for deposit *percentages*			Depth below surface (m)	*percentages*					
Fines	Sand	Gravel		Fines	Sand			Gravel	
				$-\frac{1}{16}$	$+\frac{1}{16}-\frac{1}{4}$	$+\frac{1}{4}-1$	+1–4	+4–16	+16–64
13	36	51	*0.5–3.9	16	7	19	11	23	24
			*3.9–5.0	10	3	10	22	23	32
			*5.0–6.5	10	4	8	23	22	33
			Mean	13	5	15	16	23	28

COMPOSITION

Depth below surface (m)	Percentages by weight in gravel fraction						
	Angular flint	Well-rounded flint	Quartzite	Quartz	Sandstone	Chalk	Others
0.5–6.5	59	4	5	2	4	19	7

TL 51 SW 26 **5263 1401** **Ardley End** **Block A**

Surface level (+75.6 m) +248 ft
Water not struck
152 mm percussion
July 1976

Overburden 0.5
Mineral 3.6
Waste 22.4+

LOG

Geological classification	Lithology	Thickness m	Depth m
	Soil and subsoil	0.5	0.5
Head Gravels	'Very clayey' sandy gravel Gravel: coarse and fine, angular to subangular flint with rounded to subrounded quartzite, well-rounded flint and rounded quartz Sand: medium and fine with coarse, orange-brown	3.6	4.1
Boulder Clay	Clay, orange-grey with chalk, flint and quartzite pebbles, becoming olive-grey from 5.9 m with additional pebbles of shale, siltstone and sandstone from 27.0 m. Passing down into clay, stiff, olive-black with pebbles as above below 23.0 m	22.4+	26.5

GRADING

Mean for deposit *percentages*			Depth below surface (m)	*percentages*					
Fines	Sand	Gravel		Fines	Sand			Gravel	
				$-\frac{1}{16}$	$+\frac{1}{16}-\frac{1}{4}$	$+\frac{1}{4}-1$	+1–4	+4–16	+16–64
22	45	33	0.5–1.5	25	17	18	7	13	20
			1.5–2.5	26	19	41	5	7	2
			2.5–4.1	17	8	17	12	22	24
			Mean	22	13	24	8	16	17

COMPOSITION

Depth below surface (m)	Percentages by weight in gravel fraction						
	Angular flint	Well-rounded flint	Quartzite	Quartz	Sandstone	Chalk	Others
0.5–4.1	65	6	20	6	1	0	2

TL 51 SW 27 **5233 1305** **Down Hall** **Block A**

Surface level (+78.6 m) +258 ft
Water not struck
152 mm percussion
August 1976

Overburden 0.1
Mineral 1.2
Waste 17.4+

LOG

Geological classification	Lithology	Thickness m	Depth m
	Soil	0.1	0.1
Head Gravels	'Clayey' gravel Gravel: fine and coarse, angular to subangular with well-rounded flint, rounded quartzite and quartz Sand: medium and coarse with fine, orange-grey	1.2	1.3
	Clay, very sandy, orange-brown with flint and quartzite pebbles	2.9	4.2
Boulder Clay	Clay, orange-brown with chalk, flint, shell fragment, quartzite, ironstone, siltstone and quartz pebbles, becoming brown from 5.5 m and olive-grey below 9.0 m	14.5+	18.7

GRADING

Mean for deposit *percentages*			Depth below surface (m)	*percentages*					
Fines	Sand	Gravel		Fines	Sand			Gravel	
				$-\frac{1}{16}$	$+\frac{1}{16}-\frac{1}{4}$	$+\frac{1}{4}-1$	+1–4	+4–16	+16–64
13	29	58	0.1–1.3	13	4	13	12	30	28

COMPOSITION

Depth below surface (m)	Percentages by weight in gravel fraction						
	Angular flint	Well-rounded flint	Quartzite	Quartz	Sandstone	Chalk	Others
0.1–1.3	62	11	16	8	1	0	2

TL 51 SW 28 **5232 1217** **Matching Hall** **Block A**

Surface level (+75.6 m) +248 ft
Water not struck
152 mm percussion
June 1976

Waste 18.6+

LOG

Geological classification	Lithology	Thickness m	Depth m
	Soil	0.2	0.2
Head Gravels	Clay, sandy, orange-brown with flints	0.9	1.1
	'Very clayey' sand Gravel: coarse and fine, well-rounded and angular flints Sand: coarse with medium and fine, orange-brown	0.4	1.5
Boulder Clay	Clay, mottled orange-grey and light grey with chalk and flint, pebbles, becoming olive-grey from 3.8 m with additional pebbles of shale and siltstone below 15.0 m	17.1+	18.6

TL 51 SW 29 **5293 1207** **Stone Hall** **Block A**

Surface level (+78.3 m) +257 ft
Water struck at +74.3 m
152 mm percussion
June 1976

Overburden 0.6
Mineral 3.8
Waste 14.2+

LOG

Geological classification	Lithology	Thickness m	Depth m
	Soil	0.6	0.6
Head Gravels	'Clayey' gravel Gravel: fine and coarse, angular to subangular with well-rounded flint, rounded quartzite, some quartz and sandstone Sand: medium and coarse with fine, orange-brown	3.8	4.4
Boulder Clay	Clay, olive-grey with chalk, flint, fossil fragment, and sandstone pebbles	14.2+	18.6

GRADING

Mean for deposit percentages			Depth below surface (m)	percentages						
				Fines	Sand			Gravel		
Fines	Sand	Gravel		$-\frac{1}{16}$	$+\frac{1}{16}-\frac{1}{4}$	$+\frac{1}{4}-1$	+1–4	+4–16	+16–64	+64
16	36	48	0.6–1.6	14	4	19	12	30	21	0
			1.6–2.6	14	5	19	13	32	17	0
			2.6–3.6	20	6	20	14	26	14	0
			*3.6–4.4	15	7	13	12	24	16	13
			Mean	16	5	18	13	28	17	3

COMPOSITION

Depth below surface (m)	Percentages by weight in gravel fraction						
	Angular flint	Well-rounded flint	Quartzite	Quartz	Sandstone	Chalk	Others
0.6–4.4	71	8	12	4	2	0	3

TL 51 SW 30 **5267 1116** **Brick House** **Block A**

Surface level (+80.8 m) +265 ft
Water struck at +64.1 m
152 mm percussion
July 1976

Waste 19.5
Bedrock 0.7+

LOG

Geological classification	Lithology	Thickness m	Depth m
	Soil	0.2	0.2
Boulder Clay	Clay, brown with chalk and flint pebbles, becoming olive-grey from 4.0 m. Passing down into clay, stiff, olive-black with fine pebbles of chalk, flint, sandstone and shale below 15.5 m	16.5	16.7
Glacial Sand and Gravel	Gravel Gravel: coarse and fine, angular to subangular with well-rounded flint, rounded to subrounded quartzite and quartz Sand: medium with coarse and fine quartz, light grey	2.8	19.5
London Clay	Clay, micaceous, olive-grey, top 0.4 m weathered to brown	0.7+	20.2

GRADING

Mean for deposit *percentages* Fines	Sand	Gravel	Depth below surface (m)	*percentages* Fines $-\frac{1}{16}$	Sand $+\frac{1}{16}-\frac{1}{4}$	$+\frac{1}{4}-1$	$+1-4$	Gravel $+4-16$	$+16$	$+64$
2	38	60	*16.7–17.7	2	3	39	10	19	27	0
			*17.7–18.7	1	1	19	10	32	37	0
			*18.7–19.5	1	2	23	6	32	34	2
			Mean	2	2	27	9	27	32	1

TL 51 SW 31 **5235 1031** **Loyter's Green** **Block A**

Surface level (+76.8 m) +252 ft
Water not struck
152 mm percussion
June 1976

Waste 17.7
Bedrock 0.7+

LOG

Geological classification	Lithology	Thickness m	Depth m
	Soil	0.2	0.2
Boulder Clay	Clay, brown with flint and quartz pebbles	0.6	0.8
	Clay, mottled light grey and orange-brown with chalk, flint, fossil fragment and sandstone pebbles, becoming olive-grey from 3.3 m. Passing down into clay, stiff, olive-black with mostly fine pebbles of chalk, siltstone, shale, flint, fossil fragment and quartz below 12.0 m	15.8	16.6
	Clay, very sandy, olive-brown with many flint and sandstone pebbles	1.1	17.7
London Clay	Clay, olive-grey	0.7+	18.4

TL 51 SW 32 5343 1481 Muchfield Common **Block A**

Surface level (+59.4 m) +195 ft
Water not struck
152 mm percussion
January 1977

Waste 10.5
Bedrock 0.5+

LOG

Geological classification	Lithology	Thickness m	Depth m
	Soil	0.2	0.2
Alluvium	Clay, silty and sandy, brown	1.2	1.4
	Clay, very sandy, orange-brown with flints	0.4	1.8
Boulder Clay	Clay, mottled light grey and orange-brown with chalk pebbles, becoming grey, below 3.0 m	8.7	10.5
London Clay	Clay, silty, micaceous, olive-grey	0.5+	11.0

TL 51 SW 33 5350 1115 Matching Green **Block A**

Surface level (+78.9 m) +259 ft
Water struck at +65.4 m
152 mm percussion
July 1976

Overburden 0.1
Mineral 2.8
Waste 10.6
Mineral 3.8
Waste 1.5
Mineral 3.0
Bedrock 0.7+

LOG

Geological classification	Lithology	Thickness m	Depth m
	Soil	0.1	0.1
Head Gravels	**a** 'Clayey' sandy gravel Gravel: fine and coarse, angular to subangular flint with rounded to subrounded quartzite, quartz and well-rounded flint Sand: medium with coarse and fine, orange-brown	2.8	2.9
Boulder Clay	Clay, brown with chalk and flint pebbles, becoming olive-grey from 3.4 m. Passing into clay, stiff, olive-black with chalk, flint, fossil fragment, sandstone and quartz pebbles below 10.5 m	10.0	12.9
	Clay, sandy, olive-grey with flint and quartz pebbles	0.6	13.5
Glacial Sand and Gravel	**b** Gravel Gravel: coarse and fine, angular to subangular with well-rounded flint, rounded to subrounded quartzite and quartz Sand: medium with coarse and fine quartz, orange- brown	3.8	17.3
	Clay, sandy, mottled orange-brown and brown with flints	1.5	18.8
	c Gravel Gravel: coarse and fine, angular to subangular with well-rounded flint, rounded to subrounded quartzite and quartz Sand: medium with coarse and fine, orange-brown	3.0	21.8
London Clay	Clay, micaceous, stiff, olive-grey, top weathered to brown	0.7+	22.5

GRADING

	Mean for deposit percentages			Depth below surface (m)	percentages					
					Fines	Sand			Gravel	
	Fines	Sand	Gravel		$-\frac{1}{16}$	$+\frac{1}{16}-\frac{1}{4}$	$+\frac{1}{4}-1$	+1–4	+4–16	+16–64
a	14	46	40	0.1–1.1	16	8	13	10	27	26
				1.1–2.9	12	8	33	13	20	14
				Mean	14	8	26	12	22	18
b	4	42	54	*13.5–14.5	10	4	21	13	29	23
				*14.5–15.5	1	5	43	10	13	28
				*15.5–16.5	1	4	24	9	26	36
				*16.5–17.3	2	5	17	12	31	33
				Mean	4	4	27	11	24	30
c	1	46	53	*18.8–19.8	2	5	28	11	24	30
				*19.8–20.8	1	5	41	10	22	21
				*20.8–21.8	1	2	24	13	27	33
				Mean	1	4	31	11	25	28
a+b+c	6	44	50	Mean	6	5	28	11	24	26

COMPOSITION

	Depth below surface (m)	Percentages by weight in gravel fraction						
		Angular flint	Well-rounded flint	Quartzite	Quartz	Sandstone	Chalk	Others
a	0.1–2.9	72	6	12	6	0	0	4
b+c	13.5–17.3 18.8–21.8	42	24	25	6	1	0	2
a+b+c	Mean	51	19	21	6	1	0	2

TL 51 SW 34 5385 1052 Waterman's End **Block A**

Surface level (+76.5 m) +251 ft
Water struck at +64.2 m
152 mm percussion
June 1976

Overburden 0.5
Mineral 1.2
Waste 17.2
Bedrock 1.1+

LOG

Geological classification	Lithology	Thickness m	Depth m
	Soil	0.2	0.2
Head Gravels	Clay, very sandy, orange-brown with flint and quartz pebbles	0.3	0.5
	a 'Clayey' gravel Gravel: coarse and fine, angular to subangular with well-rounded flint, rounded to subrounded quartzite and some rounded sandstone and quartz Sand: medium and coarse with fine, orange-brown	1.2	1.7
Boulder Clay	Clay, olive-grey with chalk, flint and siltstone pebbles. Passing down into clay, stiff, olive-black with fine pebbles of chalk, flint, siltstone, fossil fragment and shale below 9.0 m	10.6	12.3
Glacial Sand and Gravel	**b** Gravel Gravel: fine and coarse, angular to subangular with well-rounded flint, rounded to subrounded quartzite and quartz Sand: medium with coarse and fine, orange-grey	1.1	13.4
	Clay, olive-brown with flint, quartz and sandstone pebbles, becoming more sandy and silty with depth	3.8	17.2
	c Gravel Gravel: coarse and fine, angular to subangular with well-rounded flint, rounded to subrounded quartzite and quartz Sand: medium and fine with coarse, orange-grey	1.5	18.7
	Clay, sandy, brown with pebbles	0.2	18.9
London Clay	Clay, olive-grey, top 0.4 m weathered to orange-brown	1.1+	20.0

GRADING

	Mean for deposit *percentages*			Depth below surface (m)	*percentages*					
	Fines	Sand	Gravel		Fines	Sand			Gravel	
					$-\frac{1}{16}$	$+\frac{1}{16}-\frac{1}{4}$	$+\frac{1}{4}-1$	$+1-4$	$+4-16$	$+16-64$
a	16	37	47	0.5–1.7	16	6	21	10	21	26
b	2	49	49	*12.3–13.4	2	7	32	10	29	20
c	3	35	62	*17.2–18.7	3	13	17	5	26	36
a+b+c	7	40	53	Mean	7	9	23	8	25	28

COMPOSITION

	Depth below surface (m)	Percentages by weight in gravel fraction						
		Angular flint	Well-rounded flint	Quartzite	Quartz	Sandstone	Chalk	Others
a	0.5–1.7	78	8	8	2	3	0	1
b+c	12.3–13.4 17.2–18.7	52	14	16	15	0	0	3
a+b+c	Mean	60	12	14	11	1	0	2

TL 51 SW 35 5452 1366 Sparrows Lane **Block A**

Surface level (+64.3 m) +211 ft — Waste 18.4+
Water not struck
152 mm percussion
January 1977

LOG

Geological classification	Lithology	Thickness m	Depth m
	Made ground	0.9	0.9
Head	Clay, sandy, mottled orange-brown and light grey, with flint and chalk pebbles	1.2	2.1
Boulder Clay	Clay, mottled orange-grey and grey with chalk and flint pebbles, becoming grey with additional pebbles of siltstone below 3.8 m	16.3+	18.4

TL 51 SW 36 5460 1225 Manwood Green **Block A**

Surface level (+83.2 m) +273 ft — Waste 18.6+
Water not struck
152 mm percussion
July 1976

LOG

Geological classification	Lithology	Thickness m	Depth m
	Soil	0.2	0.2
Boulder Clay	Clay, mottled brown and light grey	0.7	0.9
	Clay, mottled brown and light grey with chalk pebbles, becoming olive-grey with additional pebbles of flint and siltstone below 4.5 m	17.7+	18.6

TL 51 SW 37 5462 1106 Stock Hall **Block A**

Surface level (+85.0 m) +279 ft
Water struck at +69.0 m
152 mm percussion
July 1976

Overburden 16.0
Mineral 6.2
Bedrock 1.0+

LOG

Geological classification	Lithology	Thickness m	Depth m
	Made ground	0.5	0.5
Boulder Clay	Silt, clayey, orange-brown with flint and chalk pebbles	1.5	2.0
	Clay, mottled olive-brown and orange-brown with flint and chalk pebbles, becoming olive-grey with additional pebbles of shale from 6.0 m, sandy silt from 9.6 to 10.5 m. Passing down into clay, stiff, olive-black with chalk, flint, shale, sandstone and siltstone pebbles below 14.7 m	14.0	16.0
Glacial Sand and Gravel	Gravel Gravel: coarse and fine, angular to subangular with well-rounded flint, rounded to subrounded quartzite and quartz Sand, medium and fine with coarse, quartz with flint, orange-brown	6.2	22.2
London Clay	Clay, stiff, grey, top 0.6 m weathered to brown	1.0+	23.2

GRADING

Mean for deposit *percentages* Fines	Sand	Gravel	Depth below surface (m)	*percentages* Fines $-\frac{1}{16}$	Sand $+\frac{1}{16}-\frac{1}{4}$	$+\frac{1}{4}-1$	$+1-4$	Gravel $+4-16$	$+16-64$
6	42	52	*16.0–17.0	16	47	14	3	9	11
			*17.0–18.0	16	41	17	4	10	12
			*18.0–19.0	4	5	34	13	20	24
			*19.0–20.0	0	2	9	6	31	52
			*20.0–21.0	1	2	12	10	25	50
			*21.0–22.2	2	5	16	10	27	40
			Mean	6	17	17	8	20	32

COMPOSITION

Depth below surface (m)	Percentages by weight in gravel fraction: Angular flint	Well-rounded flint	Quartzite	Quartz	Sandstone	Chalk	Others
16.0–22.2	45	19	23	12	0	0	1

TL 51 SW 38 5471 1021 Little Laver Hall **Block A**

Surface level (+82.3 m) +270 ft
Water struck at +70.8 m
152 mm percussion
June 1976

Overburden 10.8
Mineral 7.1
Bedrock 1.0+

LOG

Geological classification	Lithology	Thickness m	Depth m
	Soil	0.2	0.2
Boulder Clay	Clay, brown with flint and quartz pebbles	0.3	0.5
	Clay, mottled orange-grey and light grey with chalk, flint and ironstone pebbles, becoming olive-grey from 3.2 m. Passing down into clay, stiff, olive-black with chalk, flint, shale and siltstone pebbles below 8.2 m	10.3	10.8
Glacial Sand and Gravel	Gravel Gravel: coarse and fine, angular to subangular with well-rounded flint, rounded to subrounded quartzite and quartz Sand: medium with fine and coarse, orange-brown	7.1	17.9
London Clay	Clay, olive-grey, top 0.7 m weathered to brown	1.0+	18.9

GRADING

Mean for deposit *percentages*			Depth below surface (m)	*percentages*					
Fines	Sand	Gravel		Fines	Sand			Gravel	
				$-\frac{1}{16}$	$+\frac{1}{16}-\frac{1}{4}$	$+\frac{1}{4}-1$	+1–4	+4–16	+16–64
3	42	55	10.8–11.5	15	23	18	6	18	20
			*11.5–12.5	4	9	19	3	19	46
			*12.5–13.5	3	7	15	4	29	42
			*13.5–14.5	1	5	21	9	35	29
			*14.5–15.5	1	7	36	8	31	17
			*15.5–16.5	2	6	45	9	18	20
			*16.5–17.9	1	4	29	13	22	31
			Mean	3	8	26	8	25	30

TL 51 SE 30 5551 1150 Fairlands **Block A**

Surface level (+80.8 m) +265 ft
Water not struck
152 mm percussion
September 1976

Waste 18.6+

LOG

Geological classification	Lithology	Thickness m	Depth m
	Made ground	0.5	0.5
Boulder Clay	Clay, mottled orange-brown and light grey with chalk and flint pebbles, becoming olive-grey with additional pebbles of sandstone, shale and siltstone below 4.0 m	18.1+	18.6

TL 51 SE 31 5573 1030 Parker's Farm **Block A**

Surface level (+86.9 m) +285 ft
Water struck at +68.2 m
152 mm percussion
August 1976

Waste 20.9
Bedrock 1.1+

LOG

Geological classification	Lithology	Thickness m	Depth m
	Soil	0.2	0.2
Boulder Clay	Clay, mottled orange-brown and light grey with chalk and flint pebbles, becoming olive-grey with additional pebbles of shale and siltstone from 7.0 m. Passing down into clay, olive-black with fine pebbles of chalk, shale, quartz, sandstone, shell fragment and siltstone below 16.0 m	18.5	18.7
Glacial Sand and Gravel	Gravel Gravel: fine and coarse, angular to subangular with well-rounded flint and rounded quartz Sand: medium, fine and coarse, quartz and some flint, light grey	2.2	20.9
London Clay	Clay, stiff, olive-grey, top weathered to brown	1.1+	22.0

GRADING

Mean for deposit *percentages*			Depth below surface (m)	*percentages*					
Fines	Sand	Gravel		Fines	Sand			Gravel	
				$-\frac{1}{16}$	$+\frac{1}{16}-\frac{1}{4}$	$+\frac{1}{4}-1$	+1–4	+4–16	+16–64
2	38	60	*18.7–19.7	2	13	16	5	33	31
			*19.7–20.9	3	7	20	13	29	28
			Mean	2	10	18	10	31	29

TL 51 SE 32 5681 1258 Mascallsbury **Block A**

Surface level (+78.3 m) +257 ft
Water not struck
152 mm percussion
July 1976

Waste 18.6+

LOG

Geological classification	Lithology	Thickness m	Depth m
	Soil	0.2	0.2
Boulder Clay	Clay, brown with flints	1.2	1.4
	Clay, mottled orange-brown and brown with chalk and flint pebbles, becoming grey below 4.8 m	17.2+	18.6

TL 51 SE 33 5628 1099 Rookwood Hall **Block A**

Surface level (+78.9 m) +259 ft Waste 20.0+
Water not struck
152 mm percussion
December 1976

LOG

Geological classification	Lithology	Thickness m	Depth m
	Made ground	1.1	1.1
Boulder Clay	Clay, mottled light grey and orange-brown, becoming grey from 4.0 m with additional pebbles of shale and ironstone from 14.3 m. Passing down into clay, stiff, dark grey with fine pebbles of chalk, shale and flint below 18.2 m	18.9+	20.0

TL 51 SE 34 5767 1408 Norrington's Cottages **Block A**

Surface level (+68.6 m) +225 ft Waste 18.8+
Water not struck
152 mm percussion
January 1977

LOG

Geological classification	Lithology	Thickness m	Depth m
	Soil	0.4	0.4
Boulder Clay	Clay, mottled brown and light grey with chalk and flint pebbles, becoming grey from 4.3 m. Passing down into clay, stiff, dark grey with chalk, shale, flint and sandstone pebbles below 17.0 m	18.4+	18.8

TL 51 SE 35 5764 1213 Berwick Berners Hall **Block A**

Surface level (+72.5 m) +238 ft Waste 19.0+
Water not struck
152 mm percussion
December 1976

LOG

Geological classification	Lithology	Thickness m	Depth m
	Soil	0.7	0.7
Boulder Clay	Clay, mottled orange-brown and light grey with chalk and flint pebbles	0.7	1.4
	Chalk gravel, 'very clayey'	1.1	2.5
	Clay, orange-grey with chalk and flint pebbles, becoming grey from 7.2 m. Passing down into clay, stiff, olive-black with pebbles as above below 16.0 m	16.5+	19.0

TL 51 SE 36 5794 1142 Longbarns **Block A**

Surface level (+69.8 m) +229 ft
Water struck at +54.7 m
152 mm percussion
August 1976

Overburden 15.1
Mineral 6.7
Bedrock 1.7+

LOG

Geological classification	Lithology	Thickness m	Depth m
	Soil	0.1	0.1
Boulder Clay	Clay, mottled orange-brown and light grey with chalk and flint pebbles, becoming olive-grey with additional pebbles of siltstone and sandstone from 7.5 m. Passing down into clay, stiff, olive-black with fine pebbles of chalk flint, fossil fragment, siltstone and shale below 14.4 m	15.0	15.1
Glacial Sand and Gravel	Gravel Gravel: coarse and fine, angular to subangular with well-rounded flint, rounded to subrounded quartzite and quartz Sand: medium and coarse with fine, quartz with flint and a trace of chalk, light grey	6.7	21.8
London Clay	Clay, micaceous, stiff, olive-grey, top 0.3 m weathered to brown	1.7+	23.5

GRADING

Mean for deposit *percentages*			Depth below surface (m)	*percentages*					
Fines	Sand	Gravel		Fines	Sand			Gravel	
				$-\frac{1}{16}$	$+\frac{1}{16}-\frac{1}{4}$	$+\frac{1}{4}-1$	+1–4	+4–16	+16–64
4	32	64	*15.1–16.1	3	3	5	10	45	34
			*16.1–17.1	4	4	7	10	42	33
			*17.1–18.1	2	6	28	11	23	30
			*18.1–19.1	15	8	26	9	13	29
			*19.1–20.1	1	4	11	6	32	46
			*20.1–21.1	1	5	24	15	22	33
			*21.1–21.8	0	3	20	9	25	43
			Mean	4	5	17	10	29	35

COMPOSITION

Depth below surface (m)	Percentages by weight in gravel fraction						
	Angular flint	Well-rounded flint	Quartzite	Quartz	Sandstone	Chalk	Others
15.1–21.8	51	17	17	12	0	0	3

TL 51 SE 37 **5720 1043** **Woodend** **Block A**

Surface level (+70.7 m) +232 ft
Water struck at +55.1 m
152 mm percussion
August 1976

Overburden 15.6
Mineral 5.8
Bedrock 1.5+

LOG

Geological classification	Lithology	Thickness m	Depth m
	Soil	0.3	0.3
Boulder Clay	Clay, mottled brown and light grey with chalk, flint and fossil fragment pebbles, becoming olive-grey with additional pebbles of siltstone and ironstone below 6.0 m	9.1	9.4
	Silt, clayey, light grey with fine chalk pebbles	3.5	12.9
	Clay, olive-grey with chalk, flint and shale pebbles	2.7	15.6
Glacial Sand and Gravel	Gravel, with clay from 18.8 to 19.2 m Gravel: coarse and fine, angular to subangular with well-rounded flint, rounded to subrounded quartz and quartzite Sand: medium and coarse with fine, quartz with flint, orange-grey	5.8	21.4
London Clay	Clay, stiff, olive-grey, top weathered to brown	1.5+	22.9

GRADING

Mean for deposit *percentages* Fines	Sand	Gravel	Depth below surface (m)	*percentages* Fines $-\frac{1}{16}$	Sand $+\frac{1}{16}-\frac{1}{4}$	$+\frac{1}{4}-1$	$+1-4$	Gravel $+4-16$	$+16-64$
8	46	46	*15.6–16.6	6	9	29	10	23	23
			*16.6–17.6	3	17	44	8	18	10
			*17.6–18.8	13	8	30	10	10	29
			*18.8–19.2	Clay					
			*19.2–20.2	1	2	11	20	27	39
			*20.2–21.4	16	2	11	20	14	37
			Mean	8	7	25	14	18	28

TL 51 SE 38 **5837 1316** **Leaden Roding Bridge** **Block A**

Surface level (+61.3 m) +201 ft
Water not struck
152 mm percussion
September 1976

Waste 18.3+

LOG

Geological classification	Lithology	Thickness m	Depth m
	Soil	0.5	0.5
Head Gravels	Silt, sandy, mottled brown and orange-brown with flint and quartz pebbles, becoming increasingly sandy with additional pebbles of fine chalk from 1.5 m	1.7	2.2
Boulder Clay	Clay, orange-grey with chalk, flint and sandstone pebbles, becoming olive-grey from 3.4 m with additional pebbles of shale and siltstone from 7.4 m. Passing down into clay, stiff, olive-black with chalk, siltstone, fossil fragment, shale and flint pebbles below 12.8 m	16.1+	18.3

TL 51 SE 39 **5868 1219** **Nether Street** **Block A**

Surface level (+58.5 m) +192 ft
Water struck at +56.0 m
152 mm percussion
September 1976

Waste 18.6+

LOG

Geological classification	Lithology	Thickness m	Depth m
	Soil	0.4	0.4
Alluvium	Clay, mottled light grey and orange-brown	1.2	1.6
	'Clayey' gravel Gravel: coarse and fine, angular with well-rounded flint, rounded to subrounded quartzite, some quartz and sandstone Sand: medium and coarse with fine, orange-grey	0.9	2.5
Boulder Clay	Silt, sandy, orange-grey with chalk and flint pebbles, becoming clay, grey from 3.5 m with additional fine pebbles of siltstone and fossil fragment from 4.0 m. Silt, sandy olive-grey from 11.8 to 12.2 m. Passing down into clay, stiff, olive-grey with fine chalk, flint, siltstone and shale pebbles below 14.4 m	16.1+	18.6

GRADING

Mean for deposit *percentages*			Depth below surface (m)	*percentages*					
Fines	Sand	Gravel		Fines	Sand			Gravel	
				$-\frac{1}{16}$	$+\frac{1}{16}-\frac{1}{4}$	$+\frac{1}{4}-1$	$+1-4$	$+4-16$	$+16-64$
14	28	58	1.6–2.5	14	5	14	9	27	31

COMPOSITION

Depth below surface (m)	Percentages by weight in gravel fraction						
	Angular flint	Well-rounded flint	Quartzite	Quartz	Sandstone	Chalk	Others
1.6–2.5	59	19	15	3	2	0	2

TL 51 SE 40 **5888 1118** **Frayes** **Block A**

Surface level (+56.7 m) +186 ft
Water struck at +54.5 m and +52.0 m
152 mm percussion
August 1976

Overburden 1.6
Mineral 1.1
Waste 2.0
Mineral 3.0
Bedrock 0.4+

LOG

Geological classification	Lithology	Thickness m	Depth m
	Soil	0.5	0.5
Alluvium	Peat, silty, dark brown	1.1	1.6
	a 'Clayey' sandy gravel Gravel: fine and coarse, angular with well-rounded flint, rounded to subrounded quartzite, quartz and some chalk Sand: medium and coarse with fine, orange-grey	1.1	2.7
Boulder Clay	Clay, olive-grey with chalk, flint and shale pebbles, becoming olive-black below 4.3 m	2.0	4.7
Glacial Sand and Gravel	**b** Sandy gravel Gravel: fine and coarse, angular to subangular and well-rounded flint with some rounded to subrounded quartzite and quartz Sand: medium and fine with coarse, orange-grey	3.0	7.7
London Clay	Clay, stiff, olive-grey, top weathered to brown	0.4+	8.1

GRADING

	Mean for deposit *percentages*			Depth below surface (m)	*percentages*					
	Fines	Sand	Gravel		Fines	Sand			Gravel	
					$-\frac{1}{16}$	$+\frac{1}{16}-\frac{1}{4}$	$+\frac{1}{4}-1$	+1–4	+4–16	+16–64
a	13	51	36	*1.6–2.7	13	6	30	15	25	11
b	1	72	27	*4.7–5.7	1	11	30	8	28	22
				*5.7–6.7	1	32	37	5	11	14
				*6.7–7.9	3	49	40	3	2	3
				Mean	1	31	36	5	14	13
a+b	4	66	30	Mean	4	24	34	8	17	13

COMPOSITION

	Depth below surface (m)	Percentages by weight in gravel fraction						
		Angular flint	Well-rounded flint	Quartzite	Quartz	Sandstone	Chalk	Others
a	1.6–2.7	65	8	14	7	0	4	2
b	4.7–7.7	56	35	5	3	0	0	1
a+b	Mean	59	28	7	4	0	1	1

TL 51 SE 41 **5824 1048** **Longbarns** **Block A**

Surface level (+63.7 m) +209 ft
Water struck at +57.1 m
152 mm percussion
August 1976

Overburden 6.6
Mineral 7.9
Bedrock 0.9+

LOG

Geological classification	Lithology	Thickness m	Depth m
	Soil	0.3	0.3
Boulder Clay	Clay, orange-grey with chalk pebbles, passing to olive-black with additional pebbles of shell fragment, quartz, siltstone, sandstone and flint below 4.4 m	6.2	6.5
	Clay, sandy, olive-brown with many pebbles of flint and quartz	0.1	6.6
Glacial Sand and Gravel	Gravel, with 'clayey' sand seams from 9.6 to 10.6 m Gravel: coarse and fine, angular to subangular and well-rounded flint with subrounded to rounded quartzite and quartz Sand: medium with coarse and fine, quartz with flint, orange-grey	7.9	14.5
London Clay	Clay, interlaminated with silt, stiff, olive-grey, top weathered to brown	0.9+	15.4

GRADING

Mean for deposit *percentages* Fines	Sand	Gravel	Depth below surface (m)	*percentages* Fines $-\frac{1}{16}$	Sand $+\frac{1}{16}-\frac{1}{4}$	$+\frac{1}{4}-1$	$+1-4$	Gravel $+4-16$	$+16-64$
2	35	63	*6.6–7.6	2	3	20	16	30	29
			*7.6–8.6	1	3	12	9	29	46
			*8.6–9.6	2	13	26	7	24	28
			*9.6–10.6	1	14	32	6	20	27
			*10.6–11.6	4	4	11	6	28	47
			*11.6–12.6	1	5	15	13	35	31
			*12.6–13.6	2	9	14	8	30	37
			*13.6–14.5	2	5	18	9	29	37
			Mean	2	7	19	9	28	35

COMPOSITION

Depth below surface (m)	Percentages by weight in gravel fraction Angular Flint	Well-rounded flint	Quartzite	Quartz	Sandstone	Chalk	Others
6.6–14.5	51	25	16	5	0	0	3

TL 51 SE 42 **5956 1301** **Meghills** **Block A**

Surface level (+73.8 m) +242 ft
Water not struck
152 mm percussion
July 1976

Waste 18.7+

LOG

Geological classification	Lithology	Thickness m	Depth m
	Soil	0.1	0.1
Boulder Clay	Clay, mottled brown and orange-brown with chalk and flint pebbles, becoming grey below 4.7 m	18.6+	18.7

TL 51 SE 43 **5963 1184** **Garnish Hall** **Block A**

Surface level (+61.0 m) +200 ft
Water not struck
152 mm percussion
September 1976

Waste 18.6+

LOG

Geological classification	Lithology	Thickness m	Depth m
	Soil	0.2	0.2
Boulder Clay	Clay, brown, with chalk and flint pebbles, mottled light grey and orange-brown from 1.4 m, becoming olive-grey with additional pebbles of siltstone from 3.0 m, passing to dark grey below 16.5 m	18.4+	18.6

TL 51 SE 44 5925 1035 Waples Mill Farm **Block A**

Surface level (+55.8 m) +183 ft
Water struck at +53.6 m and +49.1 m
152 mm percussion
August 1976

Overburden 1.2
Mineral 2.0
Waste 3.5
Mineral 3.9
Bedrock 0.2+

LOG

Geological classification	Lithology	Thickness m	Depth m
	Soil	0.3	0.3
Alluvium	Clay, brown	0.7	1.0
	Clay, sandy, orange-brown with flints	0.2	1.2
	a 'Clayey' gravel Gravel: fine and coarse, angular with well-rounded flint, rounded to subrounded quartzite, quartz and chalk Sand: medium and coarse with fine, quartz with flint and chalk	2.0	3.2
Boulder Clay	Clay, olive-black with fine chalk, flint, siltstone, fossil fragment and sandstone pebbles	3.5	6.7
Glacial Sand and Gravel	**b** Gravel Gravel: coarse and fine, angular to subangular and well-rounded flint with some rounded to subrounded quartzite and quartz Sand: medium with coarse and fine	3.9	10.6
London Clay	Clay, stiff, olive-grey, top weathered to olive-brown	0.2+	10.8

GRADING

	Mean for deposit *percentages*			Depth below surface (m)	*percentages*					
	Fines	Sand	Gravel		Fines	Sand			Gravel	
					$-\frac{1}{16}$	$+\frac{1}{16}-\frac{1}{4}$	$+\frac{1}{4}-1$	$+1-4$	$+4-16$	$+16-64$
a	12	28	60	1.2–2.2	14	5	21	8	26	26
				*2.2–3.2	9	1	9	13	42	26
				Mean	12	3	15	10	34	26
b	4	35	61	*6.7–7.7	6	5	20	8	31	30
				*7.7–8.7	3	5	25	8	27	32
				*8.7–9.7	7	7	25	10	26	25
				*9.7–10.6	2	3	12	8	28	47
				Mean	4	5	21	9	28	33
a+b	7	32	61	Mean	7	4	19	9	30	31

COMPOSITION

	Depth below surface (m)	Percentages by weight in gravel fraction						
		Angular flint	Well-rounded flint	Quartzite	Quartz	Sandstone	Chalk	Others
a	1.2–3.2	65	10	9	5	0	6	5
b	6.7–10.6	49	44	4	2	0	0	1
a+b	Mean	54	33	6	3	0	2	2

TL 61 NW 3 **6070 1966** **Canfield Thrift** **Block B**

Surface level (+82.3 m) +270 ft
Water not struck
152 mm percussion
August 1977

Waste 15.0
Bedrock 1.5+

LOG

Geological classification	Lithology	Thickness m	Depth m
	Soil	0.3	0.3
Boulder Clay	Clay, brown with flints	0.6	0.9
	Clay, brown with chalk and flint pebbles, mottled to grey from 1.3 m, becoming grey with additional pebbles of sandstone, quartz and shale from 3.0 m, passing to olive-black below 8.0 m	12.8	13.7
	Clay, sandy, olive-brown with flint, shale, chalk, sandstone and quartz pebbles	1.3	15.0
London Clay	Clay, stiff, olive-grey	1.5+	16.5

TL 61 NW 4 **6022 1852** **Bacon End** **Block B**

Surface level (+72.8 m) +239 ft
Water struck at +53.7 m
152 mm percussion
July 1977

Overburden 16.1
Mineral 4.5
Bedrock 0.7+

LOG

Geological classification	Lithology	Thickness m	Depth m
	Soil	0.2	0.2
Boulder Clay	Clay, sandy, mottled brown and light grey with flint and quartz pebbles	2.1	2.3
	Clay, mottled orange-brown and light grey with chalk and flint pebbles, becoming olive-grey with additional pebbles of shale from 4.6 m. Passing down into clay, stiff, olive-black with fine pebbles of chalk, shale, flint, sandstone and siltstone below 10.0 m	12.2	14.5
	Clay, very sandy, olive-brown with flint, quartz and quartzite pebbles	1.6	16.1
Glacial Sand and Gravel	'Very clayey' sandy gravel Gravel: coarse and fine, angular to subangular with well-rounded flint, rounded to subrounded quartzite and quartz Sand: medium with fine and coarse, orange-grey	4.5	20.6
London Clay	Clay, stiff, olive-grey	0.7+	21.3

GRADING

Mean for deposit *percentages* Fines	Sand	Gravel	Depth below surface (m)	*percentages* Fines $-\frac{1}{16}$	Sand $+\frac{1}{16}-\frac{1}{4}$	$+\frac{1}{4}-1$	$+1-4$	Gravel $+4-16$	$+16-64$
22	58	20	*16.1–17.9	33	16	31	4	10	6
			*17.9–19.1	31	17	31	5	13	3
			*19.1–20.6	2	8	55	8	7	20
			Mean	22	13	39	6	10	10

TL 61 NW 5 6083 1767 Rands **Block B**

Surface level (+88.1 m) +289 ft
Water struck at +73.4 m
152 mm percussion
July 1977

Waste 17.9
Bedrock 0.6+

LOG

Geological classification	Lithology	Thickness m	Depth m
	Soil	0.2	0.2
Boulder Clay	Clay, sandy, brown with flints	0.4	0.6
	Clay, mottled orange-grey and light grey, containing chalk and flint pebbles with silt and chalk gravel seams, becoming olive-grey from 8.6 m. Passing down into clay, stiff, olive-black with chalk, flint, shale and quartz pebbles below 9.8 m	13.1	13.7
	Clay, sandy, olive-grey with flint, shale, quartz and quartzite pebbles	1.0	14.7
Glacial Sand and Gravel	Gravel Gravel: coarse and fine, angular to subangular with well-rounded flint, rounded to subrounded quartz and quartzite Sand: medium with coarse and fine, quartz and flint, orange-grey	3.2	17.9
London Clay	Clay, stiff, olive-grey, top weathered to brown	0.6+	18.5

GRADING

Mean for deposit *percentages*			Depth below surface (m)	*percentages*					
Fines	Sand	Gravel		Fines	Sand			Gravel	
				$-\frac{1}{16}$	$+\frac{1}{16}-\frac{1}{4}$	$+\frac{1}{4}-1$	+1–4	+4–16	+16–64
4	31	65	*14.7–16.1	8	5	27	4	23	33
			*16.1–17.9	1	4	17	7	17	54
			Mean	4	4	21	6	20	45

TL 61 NW 6 6031 1658 High Trees **Block B**

Surface level (+79.6 m) +261 ft
Water not struck
152 mm percussion
August 1977

Waste 14.3
Bedrock 0.8+

LOG

Geological classification	Lithology	Thickness m	Depth m
	Soil	0.3	0.3
Boulder Clay	Clay, mottled brown and grey with chalk and flint pebbles, clayey sand from 4.4 to 5.3 m, becoming grey with additional pebbles of quartz, sandstone and shale from 5.3 m, passing to dark grey below 11.4 m	14.0	14.3
London Clay	Clay, micaceous, olive-grey, top weathered to olive-brown	0.8+	15.1

TL 61 NW 7 6052 1536 Monk's Farm **Block B**

Surface level (+68.3 m) +224 ft
Water not struck
152 mm percussion
August 1977

Waste 18.7+

LOG

Geological classification	Lithology	Thickness m	Depth m
	Soil and subsoil	0.4	0.4
Boulder Clay	Clay, brown with chalk and flint pebbles, mottled to grey from 1.2 m, clayey sand from 3.6 to 4.0 m, becoming grey with additional pebbles of shale, sandstone and fossil fragments from 4.0 m. Passing down into clay, stiff, dark grey with pebbles as above below 10.0 m	18.3+	18.7

TL 61 NW 8 6143 1861 Philpot End **Block B**

Surface level (+82.9 m) +272 ft
Water struck at +71.4 m
152 mm percussion
July 1977

Overburden 11.5
Mineral 6.3
Bedrock 0.4+

LOG

Geological classification	Lithology	Thickness m	Depth m
	Soil	0.2	0.2
Boulder Clay	Clay, mottled brown and light grey with chalk pebbles, becoming olive-grey with additional pebbles of shale, flint and ironstone from 4.4 m. Passing down into clay, stiff, olive-black with chalk, flint, sandstone, quartzite, quartz and shale pebbles below 6.3 m	11.0	11.2
	Clay, very sandy, olive-brown with flint and quartz pebbles	0.3	11.5
Glacial Sand and Graval	Gravel, with silty sand seams from 16.0 to 17.8 m Gravel: coarse and fine, angular to subangular with well-rounded flint, rounded to well-rounded quartzite and quartz Sand: medium with coarse and fine, quartz with flint, light grey	6.3	17.8
London Clay	Clay, stiff, olive-grey, top 0.1 m weathered to brown	0.4+	18.2

GRADING

Mean for deposit *percentages*			Depth below surface (m)	*percentages*					
Fines	Sand	Gravel		Fines	Sand			Gravel	
				$-\frac{1}{16}$	$+\frac{1}{16}-\frac{1}{4}$	$+\frac{1}{4}-1$	+1–4	+4–16	+16
6	46	48	*11.5–13.0	7	2	11	9	32	39
			*13.0–14.8	5	5	29	11	23	27
			*14.8–15.9	7	4	40	13	14	22
			*15.9–17.8	4	4	47	9	15	21
			Mean	6	4	32	10	21	27

COMPOSITION

Depth below surface	Percentages by weight in gravel fraction						
	Angular flint	Well-rounded flint	Quartzite	Quartz	Sandstone	Chalk	Others
11.5–17.8	42	18	26	10	0	0	4

TL 61 NW 9 **6130 1611** **Coopers** **Block B**

Surface level (+77.4 m) +254 ft
Water struck at +71.8 m
152 mm percussion
August 1977

Overburden 5.1
Mineral 3.0
Bedrock 1.1+

LOG

Geological classification	Lithology	Thickness m	Depth m
	Soil and subsoil	0.5	0.5
Boulder Clay	Clay, brown with chalk pebbles, becoming mottled brown and light grey with additional pebbles of quartz, siltstone and fossil fragment from 1.5 m, passing to olive-black with mostly fine pebbles below 3.7 m	4.6	5.1
Glacial Sand and Gravel	'Clayey' sandy gravel, with very sandy pebbly clay from 6.5 to 7.1 m Gravel: coarse and fine, angular to subangular with well-rounded flint, rounded to subrounded quartzite, quartz and some sandstone Sand: medium with coarse and fine, orange	3.0	8.1
London Clay	Clay, stiff, olive-grey, top 0.9 m weathered to brown	1.1+	9.2

GRADING

Mean for deposit *percentages*			Depth below surface (m)	*percentages*					
Fines	Sand	Gravel		Fines	Sand			Gravel	
				$-\frac{1}{16}$	$+\frac{1}{16}-\frac{1}{4}$	$+\frac{1}{4}-1$	+1–4	+4–16	+16–64
14	60	26	*5.1–6.5	18	5	50	14	10	3
			6.5–7.1	Pebbly sandy clay					
			*7.1–8.1	9	4	36	7	12	32
			Mean	14	5	44	11	11	15

COMPOSITION

Depth below surface	Percentages by weight in gravel fraction						
	Angular flint	Well-rounded flint	Quartzite	Quartz	Sandstone	Chalk	Others
5.1–8.1	58	10	18	8	4	0	2

TL 61 NW 10 6298 1974 Martels **Block B**

Surface level (+82.9 m) +272 ft
Water struck at +69.8 m
152 mm percussion
August 1977

Waste 14.7
Bedrock 1.0+

LOG

Geological classification	Lithology	Thickness m	Depth m
	Soil	0.3	0.3
Boulder Clay	Clay, mottled brown and grey with chalk, quartzite, flint and shale pebbles, becoming grey from 5.4 m, passing to dark grey with mostly fine pebbles below 11.5 m	12.8	13.1
Glacial Sand and Gravel	Gravel	1.6	14.7
	Gravel: fine and coarse, subangular and rounded flints with quartz and quartzite		
	Sand: medium and coarse with fine, quartz with flint		
London Clay	Clay: micaceous, grey, top 0.2 m weathered to brown	1.0+	15.7

GRADING

Mean for deposit *percentages*			Depth below surface (m)	*percentages*					
Fines	Sand	Gravel		Fines	Sand			Gravel	
				$-\frac{1}{16}$	$+\frac{1}{16}-\frac{1}{4}$	$+\frac{1}{4}-1$	+1–4	+4–16	+16–64
8	39	53	*13.1–14.7	8	3	22	14	30	23

TL 61 NW 11 6256 1853 Roffey **Block B**

Surface level (+78.0 m) +256 ft
Water struck at +72.5 m
152 mm percussion
August 1977

Overburden 5.5
Mineral 2.8
Bedrock 0.8+

LOG

Geological classification	Lithology	Thickness m	Depth m
	Soil and subsoil	0.6	0.6
Boulder Clay	Clay, brown with flints	0.3	0.9
	Clay, brown with chalk and flint pebbles, becoming olive-grey with additional pebbles of shale and fossil fragments. Passing down into clay, stiff, olive-black with pebbles as above below 2.8 m	4.6	5.5
Glacial Sand and Gravel	Gravel Gravel: coarse and fine, angular to subangular with well-rounded flint, rounded to subrounded quartzite and quartz Sand: medium and coarse with fine, quartz with flint	2.8	8.3
London Clay	Clay, stiff, olive-grey, top weathered to brown	0.8+	9.1

GRADING

Mean for deposit *percentages*			Depth below surface (m)	*percentages*					
Fines	Sand	Gravel		Fines	Sand			Gravel	
				$-\frac{1}{16}$	$+\frac{1}{16}-\frac{1}{4}$	$+\frac{1}{4}-1$	+1–4	+4–16	+16–64
2	38	60	*5.5–6.5	1	2	16	10	29	42
			*6.5–7.5	3	4	29	17	25	22
			*7.5–8.3	1	2	17	14	29	37
			Mean	2	3	21	14	27	33

TL 61 NW 12 6201 1758 Proverbs Green **Block B**

Surface level (+82.3 m) +270 ft
Water struck at +72.3 m
152 mm percussion
August 1977

Overburden 8.8
Mineral 5.5
Bedrock 0.8+

LOG

Geological classification	Lithology	Thickness m	Depth m
	Soil	0.1	0.1
Boulder Clay	Clay, brown with flints	0.5	0.6
	Clay, mottled grey and brown with chalk and flint pebbles, passing to dark grey with additional pebbles of shale and quartz below 4.6 m	5.9	6.5
	Clay, very sandy, olive-brown with quartz and flint pebbles	2.3	8.8
Glacial Sand and Gravel	Clayey pebbly sand, with clay and silt seams from 8.8 to 11.8 m Gravel: coarse and fine, angular to subangular and well-rounded flint, rounded quartzite, some quartz and sandstone Sand: medium and fine with coarse, quartz with some flint and a trace of chalk	5.5	14.3
London Clay	Clay, grey, top 0.4 m weathered to brown	0.8+	15.1

GRADING

Mean for deposit *percentages*			Depth below surface (m)	*percentages*					
Fines	Sand	Gravel		Fines	Sand			Gravel	
				$-\frac{1}{16}$	$+\frac{1}{16}-\frac{1}{4}$	$+\frac{1}{4}-1$	+1–4	+4–16	+16–64
15	70	15	8.8–9.8	34	24	40	1	1	0
			*9.8–10.8	20	34	44	2	0	0
			*10.8–11.8	16	31	52	1	0	0
			*11.8–12.8	6	15	63	3	6	7
			*12.8–14.3	4	8	34	9	22	23
			Mean	15	21	45	4	7	8

COMPOSITION

Depth below surface (m)	Percentages by weight in gravel fraction						
	Angular flint	Well-rounded flint	Quartzite	Quartz	Sandstone	Chalk	Others
8.8–14.3	35	25	33	4	2	0	1

TL 61 NW 13 **6282 1697** **Shonks** **Block B**

Surface level (+89.3 m) +293 ft — Waste 18.7+
Water not struck
152 mm percussion
August 1977

LOG

Geological classification	Lithology	Thickness m	Depth m
	Soil	0.2	0.2
Boulder Clay	Clay, dark brown with flints	0.4	0.6
	Clay, mottled grey and brown with chalk and flint pebbles, clayey chalk gravel from 6.0 to 6.2 m, becoming grey with additional pebbles of shale, quartz and fossil fragment from 6.2 m, passing to dark grey below 14.5 m	18.1+	18.7

TL 61 NW 14 **6244 1541** **Green's Farm** **Block B**

Surface level (+74.7 m) +245 ft — Waste 19.0+
Water not struck
152 mm percussion
August 1977

LOG

Geological classification	Lithology	Thickness m	Depth m
	Soil	0.3	0.3
Boulder Clay	Clay, brown with flints	0.3	0.6
	Clay, mottled orange-grey and light grey with chalk and flint pebbles, becoming olive-grey with additional pebbles of ironstone and quartz. Passing down into clay, stiff, olive-black with fine pebbles of chalk, shale, flint, quartz and siltstone below 17.8 m	18.4+	19.0

TL 61 NW 15 **6343 1879** **Sallets Green** **Block C**

Surface level (+87.2 m) +286 ft — Waste 18.6+
Water not struck
152 mm percussion
August 1977

LOG

Geological classification	Lithology	Thickness m	Depth m
	Soil	0.2	0.2
Boulder Clay	Clay, dark brown with flints	0.4	0.6
	Clay, light brown with chalk, flint, shale, quartz and sandstone pebbles, becoming grey from 8.0 m, passing to dark grey with mostly fine pebbles below 16.2 m	18.0+	18.6

TL 61 NW 16 6338 1783 Garnetts Wood **Block B**

Surface level (+89.6 m) +294 ft
Water struck at +70.9 m
152 mm percussion
August 1977

Waste 23.7
Bedrock 0.9+

LOG

Geological classification	Lithology	Thickness m	Depth m
	Soil	0.3	0.3
Head Gravels	Clay, sandy, mottled orange and light grey with flints, becoming brown towards base	2.4	2.7
Boulder Clay	Clay, silty, brown with chalk and flint pebbles	3.0	5.7
	Silt, laminated with sand, orange-grey	2.0	7.7
	Silt, brown with many chalk pebbles	2.3	10.0
	Clay, olive-grey with chalk, shale, ironstone and flint pebbles, passing to olive-black below 16.7 m	7.2	17.2
	Clay, sandy, olive-grey with flint, quartz and siltstone pebbles	0.6	17.8
Glacial Sand and Gravel	Clay, very sandy, mottled orange-brown and light grey with flints	0.9	18.7
	'Clayey' sandy gravel Gravel: fine and coarse, angular to subangular with well-rounded flints, rounded to subrounded quartzite, quartz and some sandstone Sand: medium with coarse and fine, yellow-grey	5.0	23.7
London Clay	Clay, stiff, olive-grey, top weathered to light grey	0.9+	24.6

GRADING

Mean for deposit *percentages*			Depth below surface (m)	*percentages*					
Fines	Sand	Gravel		Fines	Sand			Gravel	
				$-\frac{1}{16}$	$+\frac{1}{16}-\frac{1}{4}$	$+\frac{1}{4}-1$	+1–4	+4–16	+16–64
12	55	33	*18.7–19.7	17	9	31	9	16	18
			*19.7–20.7	11	10	38	11	14	16
			*20.7–21.7	3	6	37	13	12	29
			*21.7–23.7	14	8	36	12	26	4
			Mean	12	8	36	11	19	14

TL 61 NW 17 6432 1872 Wellstye Green **Block C**

Surface level (+78.6 m) +258 ft
Water struck at +69.3 m
152 mm percussion
August 1977

Waste 11.7
Bedrock 0.8+

LOG

Geological classification	Lithology	Thickness m	Depth m
	Soil	0.2	0.2
Boulder Clay	Clay, mottled light grey and brown with chalk and flint pebbles, becoming olive-grey with additional pebbles of quartz, shale and sandstone from 4.2 m. Passing down into clay, stiff, olive-black with chalk, flint, shale, siltstone and quartzite pebbles below 5.4 m	9.1	9.3
Glacial Sand and Gravel	Gravel Gravel: coarse and fine, angular to subangular and well-rounded flint with rounded to subrounded quartz and quartzite Sand: medium and coarse with some fine, quartz with flint, orange-brown	2.4	11.7
London Clay	Clay, stiff, olive-grey, top weathered to olive-brown	0.8+	12.5

GRADING

Mean for deposit *percentages* Fines	Sand	Gravel	Depth below surface (m)	*percentages* Fines	Sand			Gravel	
				$-\frac{1}{16}$	$+\frac{1}{16}-\frac{1}{4}$	$+\frac{1}{4}-1$	+1–4	+4–16	+16–64
2	41	57	*9.3–10.3	2	2	31	15	23	27
			*10.3–11.7	2	2	19	15	25	37
			Mean	2	2	24	15	24	33

TL 61 NW 18 6432 1755 Pyes **Block C**

Surface level (+78.6 m) +258 ft
Water struck at +68.3 m
152 mm percussion
August 1977

Waste 11.0
Bedrock 1.2+

LOG

Geological classification	Lithology	Thickness m	Depth m
	Soil	0.2	0.2
Boulder Clay	Clay, orange-grey, with chalk and flint pebbles, becoming olive-grey from 4.0 m. Passing down into clay, stiff, olive-black with chalk, flint, quartzite, shale and quartz pebbles below 7.0 m	10.0	10.2
	Clay, very sandy, olive-brown with flint and quartz pebbles	0.1	10.3
Glacial Sand and Gravel	'Very clayey' pebbley sand Gravel: fine and coarse, well-rounded flint, rounded quartzite with quartz and angular flint Sand: medium with coarse and fine, flint and quartz, light grey	0.7	11.0
London Clay	Clay, stiff, olive-grey	1.2+	12.2

TL 61 NW 19 6408 1623 Yewtree **Block D**

Surface level (+89.0 m) +292 ft
Water not struck
152 mm percussion
July 1977

Waste 19.4+

LOG

Geological classification	Lithology	Thickness m	Depth m
	Soil	0.2	0.2
Boulder Clay	Clay, mottled brown and orange-brown with chalk and flint pebbles, becoming olive-grey with additional fine pebbles of siltstone, fossil fragment and shale from 8.0 m, passing to olive-black below 12.2 m	15.2	15.4
	Clay, very sandy, light grey with flint, quartz and quartzite pebbles, becoming 'very clayey' pebbly sand towards base	4.0+	19.4

TL 61 NE 18 6539 1963 Barnston Lodge **Block C**

Surface level (+72.2 m) +237 ft
Water struck at +64.3 m
152 mm percussion
July 1977

Overburden 5.0
Mineral 4.0
Bedrock 1.1+

LOG

Geological classification	Lithology	Thickness m	Depth m
	Made ground	0.5	0.5
Boulder Clay	Clay, brown with chalk, flint, quartz and quartzite pebbles, 'very clayey' sand from 3.7 to 4.0 m	4.5	5.0
Glacial Sand and Gravel	Sandy gravel Gravel: fine and coarse, angular to subangular with well-rounded flint, rounded to subrounded quartzite, quartz and some sandstone Sand: medium with fine and coarse, quartz with flint	4.0	9.0
London Clay	Clay, olive-grey, top 0.7 m weathered to brown	1.1+	10.1

GRADING

Mean for deposit *percentages*			Depth below surface (m)	*percentages*					
Fines	Sand	Gravel		Fines	Sand			Gravel	
				$-\frac{1}{16}$	$+\frac{1}{16}-\frac{1}{4}$	$+\frac{1}{4}-1$	$+1-4$	$+4-16$	$+16-64$
8	48	44	5.0–6.0	17	14	49	5	11	4
			6.0–7.0	8	15	26	8	23	20
			7.0–7.9	4	2	23	9	28	34
			*7.9–9.0	3	3	25	11	31	27
			Mean	8	9	31	8	23	21

COMPOSITION

Depth below surface (m)	Percentages by weight in gravel fraction						
	Angular flint	Well-rounded flint	Quartzite	Quartz	Sandstone	Chalk	Others
5.0–9.0	48	20	19	9	3	0	1

TL 61 NE 19 **6537 1834** **Hounslow Green** **Block C**

Surface level (+77.7 m) +255 ft
Water struck at +68.2 m
152 mm percussion
July 1977

Overburden 5.9
Mineral 4.6
Bedrock 1.1+

LOG

Geological classification	Lithology	Thickness m	Depth m
	Soil	0.2	0.2
Boulder Clay	Clay, sandy, light brown with flint, quartz and quartzite, pebbles, becoming grey towards base	4.7	4.9
Glacial Sand and Gravel	Clay, very sandy with flint, quartz and quartzite pebbles	1.0	5.9
	Gravel Gravel: fine and coarse, angular to subangular with well-rounded flint, rounded to subrounded quartzite, quartz and some sandstone Sand: medium with coarse and fine	4.6	10.5
London Clay	Clay, olive-grey, top 0.7 m weathered to brown	1.1+	11.6

GRADING

Mean for deposit *percentages*			Depth below surface (m)	*percentages*					
Fines	Sand	Gravel		Fines	Sand			Gravel	
				$-\frac{1}{16}$	$+\frac{1}{16}-\frac{1}{4}$	$+\frac{1}{4}-1$	+1–4	+4–16	+16–64
10	43	47	5.9–6.5	39	8	24	5	9	15
			6.5–7.5	6	5	25	10	26	28
			7.5–8.5	6	4	33	8	23	26
			*8.5–9.5	8	3	29	11	29	20
			*9.5–10.5	3	5	33	9	25	25
			Mean	10	5	29	9	24	23

COMPOSITION

Depth below surface (m)	Percentages by weight in gravel fraction						
	Angular flint	Well-rounded flint	Quartzite	Quartz	Sandstone	Chalk	Others
5.9–10.5	45	21	23	6	3	0	2

TL 61 NE 20 6528 1690 Lawn Hall **Block C**

Surface level (+86.9 m) +285 ft
Water struck at +68.3 m
150 mm percussion
July 1977

Waste 18.6+

LOG

Geological classification	Lithology	Thickness m	Depth m
	Soil	0.3	0.3
Boulder Clay	Clay, brown with flints	0.3	0.6
	Clay, mottled orange-brown and light grey with chalk and flint pebbles, becoming brown towards base	4.7	5.3
	Clay, mottled olive-brown and red-brown with shale, sandstone, limestone and chalk pebbles	0.9	6.2
	Chalk gravel, 'very clayey'	0.6	6.8
	Clay, orange-grey with chalk and flint pebbles, becoming olive-grey from 8.0 m with additional pebbles of shale and quartz towards base	8.0	14.8
	Clay, very sandy, dark brown with flint, quartz, quartzite and sandstone pebbles, becoming increasingly sandy towards base	3.8+	18.6

TL 61 NE 21 6502 1563 Rolfe's Farm **Block D**

Surface level (+83.5 m) +274 ft
Water struck at +67.0 m
152 mm percussion
July 1977

Overburden 15.9
Mineral 5.7
Bedrock 0.5+

LOG

Geological classification	Lithology	Thickness m	Depth m
	Soil	0.2	0.2
Boulder Clay	Clay, light brown with chalk pebbles, clayey silt from 0.7 to 1.2 m, becoming olive-grey with additional pebbles of shale, flint and siltstone from 6.3 m. Passing down into clay, stiff, olive-black with pebbles as above below 10.0 m	12.7	12.9
	Clay, sandy, olive-brown with flint and quartz pebbles becoming orange and increasingly sandy towards base	3.0	15.9
Glacial Sand and Gravel	Gravel	5.7	21.6
	Gravel: coarse and fine, angular to subangular and well-rounded flint with rounded to subrounded quartzite and some quartz		
	Sand: medium and coarse with fine, quartz with flint, orange		
London Clay	Clay, stiff, olive-grey, top 0.3 m weathered to brown	0.5+	22.1

GRADING

Mean for deposit *percentages*			Depth below surface (m)	*percentages*					
Fines	Sand	Gravel		Fines	Sand			Gravel	
				$-\frac{1}{16}$	$+\frac{1}{16}-\frac{1}{4}$	$+\frac{1}{4}-1$	+1–4	+4–16	+16–64
4	41	55	*15.9–16.9	7	1	14	11	24	43
			*16.9–17.9	2	2	22	15	28	31
			*17.9–18.9	3	3	22	10	28	34
			*18.9–19.9	0	3	31	12	24	30
			*19.9–21.6	5	4	35	13	20	23
			Mean	4	3	26	12	24	31

COMPOSITION

Depth below surface (m)	Percentages by weight in gravel fraction						
	Angular flint	Well-rounded flint	Quartzite	Quartz	Sandstone	Chalk	Others
15.9–21.6	45	26	21	4	1	0	3

TL 61 NE 22 6641 1961 Absolpark **Block C**

Surface level (+64.3 m) +211 ft
Water not struck
152 mm percussion
July 1977

Waste 2.3
Bedrock 2.9+

LOG

Geological classification	Lithology	Thickness m	Depth m
	Made ground	0.4	0.4
Boulder Clay	'Very clayey' sandy gravel Gravel: flint with quartzite and quartz Sand: medium and fine quartz	0.5	0.9
	Clay, brown with flints	1.4	2.3
London Clay	Clay, mottled grey and brown	2.9+	5.2

TL 61 NE 23 6664 1860 North End **Block C**

Surface level (+60.7 m) +199 ft
Water struck at +56.7 m
152 mm percussion
August 1977

Overburden 0.4
Mineral 4.2
Bedrock 1.6+

LOG

Geological classification	Lithology	Thickness m	Depth m
	Soil and subsoil	0.4	0.4
Glacial Sand and Gravel	'Very clayey' sandy gravel, becomes less 'clayey' with depth Gravel: fine and coarse, angular to subangular with well-rounded flint, rounded to subrounded quartz and quartzite Sand: medium and fine with coarse, orange-brown	4.2	4.6
London Clay	Clay, stiff, olive-grey, top 1.3 m weathered to brown	1.6+	6.2

GRADING

Mean for deposit *percentages*			Depth below surface (m)	*percentages*					
Fines	Sand	Gravel		Fines	Sand			Gravel	
				$-\frac{1}{16}$	$+\frac{1}{16}-\frac{1}{4}$	$+\frac{1}{4}-1$	+1–4	+4–16	+16–64
21	44	35	0.4–1.5	34	17	27	2	13	7
			1.5–2.6	26	8	14	7	27	18
			2.6–4.0	15	20	25	8	16	16
			*4.0–4.6	3	10	29	8	18	32
			Mean	21	15	23	6	18	17

TL 61 NE 24　**6675 1715**　**King's Farm**　　**Block C**

Surface level (+66.1 m) +217 ft
Water struck at +61.9 m
152 mm percussion
June 1977

Overburden 1.9
Mineral 5.1
Bedrock 1.0+

LOG

Geological classification	Lithology	Thickness m	Depth m
	Soil and subsoil	0.4	0.4
Glacial Sand and Gravel	Clay, very sandy, mottled brown, orange-brown and light grey with flint and quartz pebbles	1.5	1.9
	'Clayey' sandy gravel, with clay seams above 2.4 m Gravel: coarse and fine, angular to subangular and well-rounded flint with rounded to subrounded quartzite and quartz Sand: medium and coarse with fine	5.1	7.0
London Clay	Clay, stiff, olive-grey, top 0.7 m weathered to brown	1.0+	8.0

GRADING

Mean for deposit *percentages* Fines	Sand	Gravel	Depth below surface (m)	*percentages* Fines $-\frac{1}{16}$	Sand $+\frac{1}{16}-\frac{1}{4}$	$+\frac{1}{4}-1$	$+1-4$	Gravel $+4-16$	$+16-64$
13	44	43	1.9–2.4	39	9	12	3	14	23
			2.4–2.8	28	10	14	5	18	25
			2.8–4.2	11	2	19	16	24	28
			*4.2–5.5	2	4	32	15	28	19
			*5.5–7.0	10	5	40	12	13	20
			Mean	13	5	27	12	20	23

COMPOSITION

Depth below surface (m)	Percentages by weight in gravel fraction: Angular flint	Well-rounded flint	Quartzite	Quartz	Sandstone	Chalk	Others
1.9–7.0	39	28	24	7	1	0	1

TL 61 NE 25 6634 1608 Oldpark Farm **Block D**

Surface level (+83.5 m) +274 ft
Water struck at +81.6 m
152 mm percussion
June 1977

Overburden 1.9
Mineral 2.0
Waste 14.7+

LOG

Geological classification	Lithology	Thickness m	Depth m
	Soil	0.3	0.3
Boulder Clay	Clay, mottled orange-grey and brown with chalk pebbles	1.6	1.9
	'Very clayey' gravel Gravel: fine and coarse, rounded chalk with some angular flint Sand: medium and coarse with fine, orange-grey	2.0	3.9
	Silt, laminated with fine sand, orange-brown	2.9	6.8
	Clay, brown with chalk and flint pebbles, becoming grey with additional pebbles of shale, quartz and quartzite from 10.0 m. Passing down into clay, stiff olive-black with pebbles as above below 12.7 m	11.8+	18.6

GRADING

Mean for deposit *percentages*			Depth below surface (m)	*percentages*					
Fines	Sand	Gravel		Fines	Sand			Gravel	
				$-\frac{1}{16}$	$+\frac{1}{16}-\frac{1}{4}$	$+\frac{1}{4}-1$	+1–4	+4–16	+16–64
35	30	35	*1.9–3.9	35	7	12	11	19	16

TL 61 NE 26 6756 1950 Felsted Mill **Block C**

Surface level (+57.6 m) +189 ft
Water not struck
152 mm percussion
July 1977

Waste 1.8
Bedrock 3.8+

LOG

Geological classification	Lithology	Thickness m	Depth m
	Soil	0.1	0.1
Head	Clay, silty and sandy with flint and quartz pebbles	0.3	0.4
	Clay, silty, mottled grey and brown with many flint and quartz pebbles	1.4	1.8
London Clay	Clay, grey, top 3.7 m weathered to brown	3.8+	5.6

TL 61 NE 27 6737 1865 Absol Bridge **Block C**

Surface level (+40.2 m) +132 ft
Water struck at +36.2 m
152 mm percussion
July 1977

Waste 6.2
Bedrock 0.7+

LOG

Geological classification	Lithology	Thickness m	Depth m
	Soil	0.4	0.4
Alluvium	Clay, silty, mottled brown and grey	2.1	2.5
	Clay, silty and sandy, olive-brown with flint and quartz pebbles	3.7	6.2
London Clay	Clay, micaceous, olive-grey	0.7+	6.9

TL 61 NE 28 **6789 1793** **Camsix Farm** **Block C**

Surface level (+53.6 m) +176 ft
Water struck at +45.6 m
152 mm percussion
July 1977

Overburden 0.2
Mineral 10.7
Bedrock 1.9+

LOG

Geological classification	Lithology	Thickness m	Depth m
	Soil	0.2	0.2
Glacial Sand and Gravel	Gravel, with sandy clay from 3.4 to 4.0 m Gravel: fine and coarse, angular to subangular with well-rounded flint, rounded to subrounded quartzite and quartz Sand: medium and coarse with fine, orange-brown	10.7	10.9
London Clay	Clay, stiff, olive-grey, top 1.4 m weathered to brown	1.9+	12.8

GRADING

Mean for deposit *percentages*			Depth below surface (m)	*percentages*					
Fines	Sand	Gravel		Fines	Sand			Gravel	
				$-\frac{1}{16}$	$+\frac{1}{16}-\frac{1}{4}$	$+\frac{1}{4}-1$	$+1-4$	$+4-16$	$+16-64$
10	43	47	0.2–1.1	39	9	18	6	13	15
			1.1–2.1	12	4	23	6	24	31
			2.1–3.4	12	3	25	14	21	25
			3.4–4.0	Sandy clay					
			4.0–5.0	10	9	31	9	21	20
			5.0–6.0	10	7	34	9	28	12
			6.0–7.0	6	5	25	10	34	20
			7.0–8.0	4	4	35	9	30	18
			*8.0–9.0	5	4	39	9	28	15
			*9.0–10.0	7	2	20	11	32	28
			*10.0–10.9	1	2	20	22	29	26
			Mean	10	5	27	11	26	21

COMPOSITION

Depth below surface (m)	Percentages by weight in gravel fraction						
	Angular flint	Well-rounded flint	Quartzite	Quartz	Sandstone	Chalk	Others
0.2–10.9	44	19	24	9	1	0	3

TL 61 NE 29 **6767 1661** **Ford End** **Block D**

Surface level (+67.7 m) +222 ft
Water not struck
152 mm percussion
June 1977

Waste 7.9
Bedrock 1.0+

LOG

Geological classification	Lithology	Thickness m	Depth m
	Soil	0.3	0.3
Boulder Clay	Clay, brown with flints Clay, orange-brown with chalk, flint and quartz pebbles, mottled to light grey from 1.3 m, becoming brown from 2.7 m. Passing down into clay, olive-grey with chalk, flint, sandstone, siltstone and shale pebbles below 5.3 m	7.6	7.9
London Clay	Clay, stiff, olive-grey, top weathered to brown	1.0+	8.9

TL 61 NE 30 6742 1564 Rolphy Green **Block D**

Surface level (+68.0 m) +223 ft
Water struck at +52.5 m
152 mm percussion
July 1977

Overburden 14.2
Mineral 9.8+

LOG

Geological classification	Lithology	Thickness m	Depth m
	Made ground	0.2	0.2
Boulder Clay	Clay, mottled orange-grey and light grey with chalk pebbles, becoming olive-brown towards base,	3.2	3.4
	Chalk gravel, 'very clayey'	0.9	4.3
	Clay, olive-brown with fine pebbles of chalk, flint, ironstone and shale, becoming olive-grey below 6.0 m	8.6	12.9
	Clay, very sandy, brown with flint, quartz, quartzite, shale and siltstone pebbles	1.3	14.2
Glacial Sand and Gravel	Sandy gravel Gravel: fine and coarse, angular to subangular with well-rounded flint, rounded to subrounded quartzite and quartz Sand: medium with fine and coarse, quartz with flint, light grey	9.8+	24.0

GRADING

Mean for deposit percentages: Fines	Sand	Gravel	Depth below surface (m)	*percentages* Fines $-\frac{1}{16}$	Sand $+\frac{1}{16}-\frac{1}{4}$	$+\frac{1}{4}-1$	$+1-4$	Gravel $+4-16$	$+16-64$
8	57	35	14.2–15.2	[16	7	17	8	37	15]
			*15.2–16.2	11	14	34	16	18	7
			*16.2–17.2	8	4	18	20	30	20
			*17.2–18.2	7	5	21	20	25	22
			*18.2–19.2	9	11	30	11	19	20
			*19.2–20.2	14	30	42	5	5	4
			*20.2–21.2	2	13	44	8	19	14
			*21.2–22.2	1	13	42	10	17	17
			*22.2–24.0	3	9	60	6	8	14
			Mean	8	12	34	11	20	15

TL 61 NE 31 **6853 1975** **Cobler's Green** **Block C**

Surface level (+67.1 m) +220 ft
Water struck at +56.9 m
152 mm percussion
August 1977

Overburden 10.2
Mineral 3.9
Bedrock 1.0+

LOG

Geological classification	Lithology	Thickness m	Depth m
	Soil	0.2	0.2
Boulder Clay	Clay, orange-brown with quartz and flint pebbles	0.3	0.5
	Clay, mottled brown and grey with chalk and flint pebbles, becoming olive-brown with additional pebbles of quartz and shale from 3.8 m, passing to dark grey with mostly fine pebbles below 4.9 m	8.6	9.1
	Clay, sandy, brown with fine pebbles of quartz and flint	1.1	10.2
Glacial Sand and Gravel	Gravel, with clay seams Gravel: coarse and fine, angular to subangular with well-rounded flint, rounded to subrounded quartzite and quartz Sand: medium with coarse and fine	3.9	14.1
London Clay	Clay, micaceous, grey, top weathered to brown	1.0+	15.1

GRADING

Mean for deposit *percentages*			Depth below surface (m)	*percentages*					
Fines	Sand	Gravel		Fines	Sand			Gravel	
				$-\frac{1}{16}$	$+\frac{1}{16}-\frac{1}{4}$	$+\frac{1}{4}-1$	+1–4	+4–16	+16–64
4	46	50	*10.2–11.2	7	4	33	6	24	26
			*11.2–12.2	2	6	29	5	24	34
			*12.2–13.2	4	7	37	9	21	22
			*13.2–14.1	2	7	34	7	15	35
			Mean	4	6	33	7	21	29

COMPOSITION

Depth below surface (m)	Percentages by weight in gravel fraction						
	Angular flint	Well-rounded flint	Quartzite	Quartz	Sandstone	Chalk	Others
10.2–14.1	48	16	25	9	0	0	2

TL 61 NE 32 6825 1895 Glandfield's Farm **Block C**

Surface level (+72.5 m) +238 ft
Water struck at +59.2 m
152 mm percussion
July 1977

Waste 18.2
Bedrock 0.3+

LOG

Geological classification	Lithology	Thickness m	Depth m
	Soil	0.2	0.2
Boulder Clay	Clay, sandy, orange-brown with flint and quartz pebbles	1.9	2.1
	Clay, brown with chalk and flint pebbles, becoming olive-grey with additional pebbles of shale and quartz below 7.8 m	11.2	13.3
Glacial Sand and Gravel	'Clayey' sandy gravel Gravel: coarse and fine, angular to subangular flint Sand: medium with coarse and fine	1.9	15.2
Boulder Clay	Clay, brown with chalk and flint pebbles. Passing down into clay, stiff, olive-grey with fine pebbles of chalk, shale, sandstone, siltstone and flint below 16.5 m	2.3	17.5
	Clay, sandy, olive-brown with flint and quartzite pebbles	0.7	18.2
London Clay	Clay, stiff, olive-brown	0.3+	18.5

GRADING

Mean for deposit percentages			Depth below surface (m)	percentages					
Fines	Sand	Gravel		Fines	Sand			Gravel	
				$-\frac{1}{16}$	$+\frac{1}{16}-\frac{1}{4}$	$+\frac{1}{4}-1$	+1–4	+4–16	+16–64
13	61	26	*13.3–14.3	8	5	32	9	18	28
			*14.3–15.2	19	10	58	10	2	1
			Mean	13	7	44	10	11	15

TL 61 NE 33 6870 1818 Prior's Green **Block C**

Surface level (+64.3 m) +211 ft
Water struck at +58.9 m
152 mm percussion
July 1977

Waste 5.9
Bedrock 2.1+

LOG

Geological classification	Lithology	Thickness m	Depth m
	Soil	0.3	0.3
Boulder Clay	Clay, light brown with flints, becoming mottled brown and grey with additional pebbles of chalk below 3.0 m	5.1	5.4
Glacial Sand and Gravel	Gravel Gravel: fine with coarse, angular to rounded flints with subrounded sandstone, quartzite and quartz Sand: medium with coarse and fine, quartz with flint	0.5	5.9
London Clay	Clay, silty, grey	2.1+	8.0

TL 61 NE 34 6859 1642 Great Appletrees **Block D**

Surface level (+65.2 m) +214 ft
Water struck at +57.2 m
152 mm percussion
June 1977

Overburden 7.8
Mineral 3.2
Bedrock 0.9+

LOG

Geological classification	Lithology	Thickness m	Depth m
	Soil	0.3	0.3
Boulder Clay	Clay, brown with chalk and flint pebbles, becoming olive-brown below 6.7 m	7.3	7.6
	Clay, sandy, red-brown with flint and quartz pebbles	0.2	7.8
Glacial Sand and Gravel	Sandy gravel Gravel: fine and coarse, angular to subangular flint and rounded quartzite with well-rounded flint and some rounded quartz and sandstone Sand: medium and coarse with fine	3.2	11.0
London Clay	Clay, stiff, olive-grey, top 0.6 m weathered to brown	0.9+	11.9

GRADING

Mean for deposit *percentages*			Depth below surface (m)	*percentages*					
Fines	Sand	Gravel		Fines	Sand			Gravel	
				$-\frac{1}{16}$	$+\frac{1}{16}-\frac{1}{4}$	$+\frac{1}{4}-1$	+1–4	+4–16	+16–64
9	53	38	*7.8–9.5	10	4	19	14	27	26
			*9.5–11.0	7	11	49	11	13	9
			Mean	9	7	33	13	20	18

COMPOSITION

Depth below surface (m)	Percentages by weight in gravel fraction						
	Angular flint	Well-rounded flint	Quartzite	Quartz	Sandstone	Chalk	Others
7.8–11.0	42	19	32	3	3	0	1

TL 61 NE 35 6837 1550 The Bucket **Block D**

Surface level (+61.3 m) +201 ft
Water struck at +48.2 m
152 mm percussion
June 1977

Overburden 13.1
Mineral 8.4
Bedrock 0.7+

LOG

Geological classification	Lithology	Thickness m	Depth m
	Soil	0.2	0.2
Boulder Clay	Clay, mottled orange-brown and light grey with chalk and flint pebbles, becoming olive-grey with additional pebbles of quartz, siltstone and shale from 6.2 m. Passing down into clay, sandy, stiff, olive-grey with flint quartz, sandstone and shale pebbles below 10.9 m	12.9	13.1
Glacial Sand and Gravel	Gravel Gravel: coarse and fine, angular to subangular with well-rounded flint, rounded to subrounded quartzite and quartz Sand: medium and coarse with fine, quartz with flint	8.4	21.5
London Clay	Clay, stiff, olive-grey	0.7+	22.2

GRADING

Mean for deposit percentages			Depth below surface (m)	percentages					
Fines	Sand	Gravel		Fines	Sand			Gravel	
				$-\frac{1}{16}$	$+\frac{1}{16}-\frac{1}{4}$	$+\frac{1}{4}-1$	+1–4	+4–16	+16–64
3	40	57	*13.1–14.1	2	2	8	13	33	42
			*14.1–15.1	1	4	24	22	26	23
			*15.1–16.1	5	3	22	21	32	17
			*16.1–17.1	3	3	23	13	26	32
			*17.1–19.3	3	4	20	18	25	30
			*19.3–20.6	2	4	23	14	25	32
			*20.6–21.5	4	4	31	9	22	30
			Mean	3	3	21	16	27	30

COMPOSITION

Depth below surface (m)	Percentages by weight in gravel fraction						
	Angular flint	Well-rounded flint	Quartzite	Quartz	Sandstone	Chalk	Others
13.1–21.5	46	18	22	9	1	0	4

TL 61 NE 36 **6957 1940** **Pondpark Farm** **Block C**

Surface level (+63.4 m) +208 ft
Water struck at +53.9 m
152 mm percussion
August 1977

Waste 12.3
Bedrock 1.0+

LOG

Geological classification	Lithology	Thickness m	Depth m
	Soil	0.1	0.1
Boulder Clay	Clay, brown with flint, limestone, quartz and sandstone pebbles	0.8	0.9
	Clay, mottled brown and grey with chalk and flint pebbles, becoming dark grey with additional pebbles of quartz, shale and sandstone below 4.9 m	8.6	9.5
Glacial Sand and Gravel	'Clayey' sandy gravel Gravel: fine and coarse, angular flint with subrounded quartz, sandstone and a trace of chalk and shale Sand: medium with coarse and fine, quartz with some chalk and shale	2.1	11.6
	Clay, brown with quartz, sandstone, flint and chalk pebbles	0.7	12.3
London Clay	Clay, micaceous, olive-brown, top weathered to brown	1.0+	13.3

GRADING

Mean for deposit *percentages*			Depth below surface (m)	*percentages*					
Fines	Sand	Gravel		Fines	Sand			Gravel	
				$-\frac{1}{16}$	$+\frac{1}{16}-\frac{1}{4}$	$+\frac{1}{4}-1$	+1–4	+4–16	+16–64
12	59	29	*9.5–10.5	20	9	32	9	19	11
			*10.5–11.6	4	8	50	10	15	13
			Mean	12	8	41	10	17	12

TL 61 NE 37 **6969 1792** **Littleypark** **Block C**

Surface level (+68.6 m) +225 ft
Water not struck
152 mm percussion
July 1977

Waste 9.4
Bedrock 1.2+

LOG

Geological classification	Lithology	Thickness m	Depth m
	Soil	0.3	0.3
Boulder Clay	Clay, sandy, mottled orange-brown and red-brown with flint and quartz pebbles	2.4	2.7
	Clay, mottled brown and light grey with chalk and flint pebbles. Passing down into clay, very sandy, brown with flint and quartz pebbles below 4.0 m	6.7	9.4
London Clay	Clay, stiff, olive-grey, top weathered to brown	1.2+	10.6

TL 61 NE 38 **6963 1672** **Littley Green** **Block D**

Surface level (+36.3 m) +119 ft
Water not struck
51 mm hand auger
July 1977

Waste 1.6
Bedrock 0.5+

LOG

Geological classification	Lithology	Thickness m	Depth m
Alluvium	Silt, sandy, brown with flints below 0.6 m	1.6	1.6
London Clay	Clay, stiff, brown	0.5+	2.1

TL 61 NE 39 **6974 1675** **Littley Green** **Block D**

Surface level (+44.8 m) +147 ft
Water not struck
51 mm hand auger
June 1977

Waste 0.6
Bedrock 1.7+

LOG

Geological classification	Lithology	Thickness m	Depth m
	Soil	0.1	0.1
Head	Clay, sandy, brown with flints	0.5	0.6
London Clay	Clay, brown	1.7+	2.3

TL 61 SW 13 **6027 1110** **Marks Hall** **Block A**

Surface level (+68.3 m) +224 ft)
Water not struck
152 mm percussion
June 1977

Waste 18.3+

LOG

Geological classification	Lithology	Thickness m	Depth m
	Soil	0.4	0.4
Boulder Clay	Clay, mottled brown and light grey with chalk and flint pebbles, becoming grey below 5.0 m	17.9+	18.3

TL 61 SW 14 6160 1477 Slough Bridge **Block B**

Surface level (+61.0 m) +200 ft
Water struck at +42.7 m
152 mm percussion
June 1977

Waste 18.3+

LOG

Geological classification	Lithology	Thickness m	Depth m
	Soil	0.2	0.2
Boulder Clay	Clay, sandy and silty, brown with flints	0.8	1.0
	Gravel Gravel: fine angular flint with some quartz Sand: coarse with some medium	0.2	1.2
	Clay, mottled grey and light grey with chalk and flint pebbles, becoming grey with additional pebbles of siltstone, quartz and shale below 3.5 m	14.2	15.4
	Clay, sandy, light grey with flint, quartz, quartzite and some chalk pebbles	2.9+	18.3

TL 61 SW 15 6201 1225 Gurtons Farm **Block E**

Surface level (+52.7 m) +173 ft
Water struck at +39.6 m
152 mm percussion
June 1977

Waste 16.7
Bedrock 0.6+

LOG

Geological classification	Lithology	Thickness m	Depth m
	Soil	0.3	0.3
Boulder Clay	Clay, brown with flints	0.5	0.8
	Clay, mottled orange-grey and light grey with chalk and flint pebbles, becoming grey from 5.0 m. Passing down into clay, stiff, olive-black with chalk, flint, quartz, shale and siltstone pebbles below 8.5 m	12.3	13.1
Glacial Sand and Gravel	Sandy gravel Gravel: fine and coarse, angular to subangular with well-rounded flint, rounded to subrounded quartzite and some quartz Sand: medium with coarse and fine, light grey	3.6	16.7
London Clay	Clay, stiff, olive-grey	0.6+	17.3

GRADING

Mean for deposit *percentages*			Depth below surface (m)	*percentages*					
Fines	Sand	Gravel		Fines	Sand			Gravel	
				$-\frac{1}{16}$	$+\frac{1}{16}-\frac{1}{4}$	$+\frac{1}{4}-1$	+1–4	+4–16	+16–64
5	66	29	*13.1–14.1	4	1	33	19	23	20
			*14.1–15.3	6	6	48	13	15	12
			*13.3–16.7	6	4	58	12	12	8
			Mean	5	4	48	14	16	13

COMPOSITION

Depth below surface (m)	Percentages by weight in gravel fraction						
	Angular flint	Well-rounded flint	Quartzite	Quartz	Sandstone	Chalk	Others
13.1–16.7	72	8	17	2	1	0	0

TL 61 SW 16 6388 1485 Stagden Cross **Block D**

Surface level (+86.9 m) +285 ft Waste 18.3+
Water not struck
152 mm percussion
July 1977

LOG

Geological classification	Lithology	Thickness m	Depth m
	Soil	0.2	0.2
Boulder Clay	Clay, brown with flints	0.9	1.1
	Clay, orange-brown with chalk and flint pebbles	4.4	5.5
	Salt, laminated, brown with some flints	2.2	7.7
	Clay, olive-brown with chalk, sandstone and siltstone pebbles	0.4	8.1
	Chalk gravel, 'very clayey'	0.6	8.7
	Clay, grey with chalk sandstone and siltstone pebbles. Passing down into clay, stiff, olive-black with fine pebbles of chalk, flint, quartzite, quartz, siltstone and shale from 14.5 m, becoming sandy, olive-brown towards base	9.6+	18.3

TL 61 SW 17 6385 1126 Great Newarks **Block E**

Surface level (+43.9 m) +144 ft Waste 19.5+
Water not struck
152 mm percussion
July 1977

LOG

Geological classification	Lithology	Thickness m	Depth m
	Soil	0.2	0.2
Boulder Clay	Clay, brown with chalk and flint pebbles, mottled to light grey from 0.8 m, becoming grey with additional pebbles of shale, fossil fragment, and sandstone below 3.0 m	19.3+	19.5

TL 61 SW 18 6311 1053 Chalk End **Block E**

Surface level (+58.8 m) +193 ft Waste 18.1+
Water struck at +40.8 m
152 mm percussion
June 1977

LOG

Geological classification	Lithology	Thickness m	Depth m
	Soil	0.3	0.3
Boulder Clay	Clay, brown with some flints	1.1	1.4
	Clay, orange-grey with chalk and flint pebbles, becoming grey with additional pebbles of shale and sandstone from 5.8 m. Sandy chalky gravel at base	16.7+	18.1

TL 61 SE 3 **6514 1444** **Wheats** **Block D**

Surface level (+82.3 m) +270 ft
Water not struck
152 mm percussion
July 1977

Waste 18.7+

LOG

Geological classification	Lithology	Thickness m	Depth m
	Soil	0.2	0.2
Boulder Clay	Clay, brown with flints	0.6	0.8
	Clay, mottled orange-grey and light grey with chalk and flint pebbles, becoming olive-brown below 4.2 m	6.4	7.2
	Chalk gravel with chalk sand and silt	1.0	8.2
	Clay, olive-grey with chalk, shale and flint pebbles	10.5+	18.7

TL 61 SE 4 **6517 1139** **Mashbury Hall** **Block E**

Surface level (+52.1 m) +171 ft
Water not struck
152 mm percussion
May 1977

Waste 18.4+

LOG

Geological classification	Lithology	Thickness m	Depth m
	Soil	0.4	0.4
Boulder Clay	Clay, mottled orange-grey and light grey with chalk pebbles, becoming grey from 3.6 m, chalk gravel from 10.3 to 10.5 m, passing to olive-grey with additional pebbles of flint and shale towards base	18.0+	18.4

TL 61 SE 5 **6502 1018** **Little Boyton Hall** **Block E**

Surface level (+59.4 m) +195 ft
Water not struck
152 mm percussion
May 1977

Waste 18.4+

LOG

Geological classification	Lithology	Thickness m	Depth m
	Soil	0.3	0.3
Boulder Clay	Clay, mottled brown and light grey with chalk pebbles, becoming grey with additional pebbles of flint below 6.7 m	18.1+	18.4

TL 61 SE 6 6603 1498 Pleshey Grange **Block D**

Surface level (+64.9 m) +213 ft
Water struck at +49.9 m
152 mm percussion
July 1977

Overburden 13.2
Mineral 9.2
Bedrock 0.5+

LOG

Geological classification	Lithology	Thickness m	Depth m
	Soil	0.3	0.3
Boulder Clay	Silt, clayey, orange-brown with flints	0.6	0.9
	Clay, mottled brown and light grey with chalk flint and shale pebbles, becoming grey with additional pebbles of sandstone and siltstone below 4.0 m	7.0	7.9
	Clay, stiff, olive-black with 0.1 m sandy seams and pebbles of chalk, flint, shale, quartz and ironstone, becoming sandy below 9.3 m	5.3	13.2
Glacial Sand and Gravel	Gravel, with clay and 'clayey' seams above 15.0 m Gravel: fine and coarse, angular to subangular with well-rounded flint, rounded to subrounded quartzite and quartz Sand: medium and coarse with fine, quartz with flint, orange-brown	9.2	22.4
London Clay	Clay, stiff olive-grey, top 0.2 m weathered to brown	0.5+	22.9

GRADING

Mean for deposit *percentages*			Depth below surface (m)	*percentages*					
Fines	Sand	Gravel		Fines	Sand			Gravel	
				$-\frac{1}{16}$	$+\frac{1}{16}-\frac{1}{4}$	$+\frac{1}{4}-1$	$+1-4$	$+4-16$	$+16-64$
3	40	57	13.2–15.0	9	7	23	13	24	24
			*15.0–16.0	2	12	23	11	29	23
			*16.0–17.0	4	6	24	12	31	23
			*17.0–18.0	3	3	38	11	19	26
			*18.0–19.0	2	1	18	14	35	30
			*19.0–20.2	1	1	13	19	40	26
			*20.2–21.2	2	2	21	12	35	28
			*21.2–22.4	1	4	16	12	26	41
			Mean	3	5	22	13	29	28

COMPOSITION

Depth below surface (m)	Percentages by weight in gravel fraction						
	Angular flint	Well-rounded flint	Quartzite	Quartz	Sandstone	Chalk	Others
13.2–22.4	50	18	22	8	0	0	2

TL 61 SE 7 6627 1402 Pleshey **Block D**

Surface level (+69.5 m) +228 ft Waste 18.9+
Water not struck
152 mm percussion
July 1977

LOG

Geological classification	Lithology	Thickness m	Depth m
	Soil	0.2	0.2
Boulder Clay	Clay, brown with chalk and flint pebbles, becoming grey with additional pebbles of flint, siltstone and shale from 3.0 m. Passing down into clay, stiff, olive-black with pebbles as above below 16.0 m	16.4	16.6
	Clay, sandy, olive-brown with flint, quartz and quartzite pebbles, becoming more sandy with depth. Passing down into sand, very clayey, olive-brown with pebbles as above below 18.7 m	2.3+	18.9

TL 61 SE 8 6655 1306 Bards Hall **Block D**

Surface level (+73.5 m) +241 ft Waste 19.0+
Water not struck
152 mm percussion
July 1977

LOG

Geological classification	Lithology	Thickness m	Depth m
	Soil	0.2	0.2
Boulder Clay	Clay, mottled orange-brown and light grey with chalk, flint and sandstone pebbles, becoming olive-grey with additional pebbles of ironstone, shale, fossil fragment and siltstone below 8.8 m	18.8+	19.0

TL 61 SE 9 6651 1184 Chignall Smealy **Block E**

Surface level (+53.0 m) +174 ft Waste 18.4+
Water not struck
152 mm percussion
May 1977

LOG

Geological classification	Lithology	Thickness m	Depth m
	Soil	0.4	0.4
Boulder Clay	Clay, light grey with some fine chalk pebbles	2.3	2.7
	Clay, silty, olive-grey with wood, peat and fragile shells	1.3	4.0
	Clay, grey with chalk, flint, fossil fragment and shale pebbles	14.4+	18.4

TL 61 SE 10 6724 1465 Dropshorts **Block D**

Surface level (+64.0 m) +210 ft
Water struck at +47.8 m
152 mm percussion
July 1977

Overburden 16.2
Mineral 7.0
Bedrock 0.8+

LOG

Geological classification	Lithology	Thickness m	Depth m
	Soil	0.2	0.2
Boulder Clay	Clay, brown with chalk pebbles, becoming olive-grey with additional pebbles of shale, fossil fragment, flint, quartz and ironstone below 7.4 m	14.7	14.9
	Clay, very sandy, olive-brown with flint, quartz and quartzite pebbles, becoming more sandy towards base	1.3	16.2
Glacial Sand and Gravel	Gravel with 'clayey' seams above 21.6 m Gravel: coarse and fine, well-rounded and angular to subangular flint, rounded to subrounded quartzite and quartz Sand: medium and coarse with fine, quartz with flint	7.0	23.2
London Clay	Clay, stiff, olive-grey	0.8+	24.0

GRADING

Mean for deposit *percentages* Fines	Sand	Gravel	Depth below surface (m)	*percentages* Fines $-\frac{1}{16}$	Sand $+\frac{1}{16}-\frac{1}{4}$	$+\frac{1}{4}-1$	$+1-4$	Gravel $+4-16$	$+16-64$
2	44	54	*16.2–17.2	3	1	16	16	22	42
			*17.2–18.2	2	2	13	12	39	32
			*18.2–19.9	2	6	47	10	16	19
			*19.9–20.9	1	5	27	9	35	23
			*20.9–21.9	2	3	25	16	22	32
			*21.9–23.2	3	6	22	15	28	26
			Mean	2	4	27	13	26	28

TL 61 SE 11 6754 1368 Fitzjohn's Farm **Block D**

Surface level (+63.4 m) +208 ft
Water struck at +46.6 m
152 mm percussion
July 1977

Overburden 16.7
Mineral 5.6
Bedrock 0.8+

LOG

Geological classification	Lithology	Thickness m	Depth m
	Made ground	0.4	0.4
Boulder Clay	Clay, orange-grey with chalk and flint pebbles, mottled to light grey from 2.5 m, becoming olive-grey with additional pebbles of quartz below 10.3 m	13.4	13.8
	Clay, very sandy, brown with flint and quartz pebbles	2.9	16.7
Glacial Sand and Gravel	Gravel, with clay seams above 19.7 m Gravel: coarse and fine, angular to subangular with well-rounded flint, rounded to subrounded quartzite and quartz Sand: medium and coarse with fine, quartz with flint, orange-brown	5.6	22.3
London Clay	Clay, stiff, olive-grey	0.8+	23.1

GRADING

Mean for deposit *percentages* Fines	Sand	Gravel	Depth below surface (m)	*percentages* Fines $-\frac{1}{16}$	Sand $+\frac{1}{16}-\frac{1}{4}$	$+\frac{1}{4}-1$	$+1-4$	Gravel $+4-16$	$+16-64$
4	31	65	*16.7–18.7	4	3	9	16	35	33
			*18.7–19.7	9	3	17	13	35	23
			*19.7–21.2	2	3	28	8	27	32
			*21.2–22.3	1	3	14	11	30	41
			Mean	4	3	16	12	32	33

COMPOSITION

Depth below surface (m)	Percentages by weight in gravel fraction: Angular flint	Well-rounded flint	Quartzite	Quartz	Sandstone	Chalk	Others
16.7–22.3	55	19	18	5	1	0	2

TL 61 SE 12 6750 1241 Blatche's Wood **Block E**

Surface level (+61.9 m) +203 ft
Water not struck
152 mm percussion
June 1977

Waste 19.5+

LOG

Geological classification	Lithology	Thickness m	Depth m
	Made ground	0.4	0.4
Boulder Clay	Clay, mottled orange-brown and light grey with chalk pebbles	0.5	0.9
	Silt, sandy, mottled orange-grey and light grey with chalk pellet seams	1.2	2.1
	Chalk gravel, 'very clayey'	0.9	3.0
	Clay, brown with chalk and flint pebbles, becoming grey with additional pebbles of shale and quartz from 6.3 m. Passing down into clay, sandy, stiff, dark grey with pebbles as above below 18.0 m	16.4	19.4
	Clay, very sandy, dark brown with flint and quartz pebbles	0.1+	19.5

TL 61 SE 13 **6767 1124** **Maple View** **Block E**

Surface level (+52.4 m) +172 ft
Water struck at +44.9 m
152 mm percussion
May 1977

Overburden 4.6
Mineral 8.0
Bedrock 0.6+

LOG

Geological classification	Lithology	Thickness m	Depth m
	Soil	0.3	0.3
Boulder Clay	Clay, brown with flints	0.6	0.9
	Clay, brown with chalk and flint pebbles, mottled to light grey from 1.9 m, becoming light grey towards base	3.7	4.6
Glacial Sand and Gravel	Gravel, with 1 cm 'clayey' seams above 6.2 m Gravel: fine and coarse, angular to subangular with well-rounded flint, rounded to subrounded quartzite and quartz Sand: medium and coarse with fine, orange-brown	8.0	12.6
London Clay	Clay, stiff, dark brown	0.6+	13.2

GRADING

Mean for deposit *percentages* Fines	Sand	Gravel	Depth below surface (m)	*percentages* Fines $-\frac{1}{16}$	Sand $+\frac{1}{16}-\frac{1}{4}$	$+\frac{1}{4}-1$	$+1-4$	Gravel $+4-16$	$+16-64$
7	43	50	4.6–5.9	15	15	40	4	14	12
			5.9–6.9	19	8	25	10	26	12
			*6.9–8.5	3	6	20	13	38	20
			*8.5–11.8	3	6	19	12	33	27
			*11.8–12.6	4	3	20	22	24	27
			Mean	7	7	24	12	29	21

COMPOSITION

Depth below surface (m)	Percentages by weight in gravel fraction: Angular flint	Well-rounded flint	Quartzite	Quartz	Sandstone	Chalk	Others
4.6–12.6	57	18	15	7	1	0	2

TL 61 SE 14 **6716 1022** **Chignall Hall** **Block E**

Surface level (+46.3 m) +152 ft
Water not struck
152 mm percussion
June 1977

Waste 18.3+

LOG

Geological classification	Lithology	Thickness m	Depth m
	Soil	0.3	0.3
Boulder Clay	Clay, mottled orange-brown and light grey with chalk and flint pebbles, becoming olive-grey with additional pebbles of quartzite and quartz below 4.5 m	11.9	12.2
	Clay, sandy, olive-brown with flint, quartzite and quartz pebbles	6.1+	18.3

TL 61 SE 15 6836 1446 Waltham Bury **Block D**

Surface level (+59.1 m) +194 ft
Water struck at +49.2 m
152 mm percussion
June 1977

Overburden 8.6
Mineral 6.3
Bedrock 0.8+

LOG

Geological classification	Lithology	Thickness m	Depth m
	Soil	0.3	0.3
Boulder Clay	Clay, brown with flints	1.5	1.8
	Clay, brown with chalk and flint pebbles, mottled to light grey from 4.7 m, passing to olive-grey with additional pebbles of ironstone and siltstone below 6.7 m	6.8	8.6
Glacial Sand and Gravel	'Clayey' sandy gravel, with clay seams above 10.6 m Gravel: fine and coarse, angular to subangular with well-rounded flint, rounded to subrounded quartz and quartzite Sand: medium and fine with coarse, quartz with flint, orange-grey	6.3	14.9
London Clay	Clay, stiff, micaceous, olive-grey	0.8+	15.7

GRADING

Mean for deposit percentages: Fines	Sand	Gravel	Depth below surface (m)	*percentages* Fines $-\frac{1}{16}$	Sand $+\frac{1}{16}-\frac{1}{4}$	Sand $+\frac{1}{4}-1$	Sand $+1-4$	Gravel $+4-16$	Gravel $+16-64$
11	65	24	8.6–9.2	25	10	13	11	30	11
			*9.2–10.3	11	14	48	5	17	5
			*10.3–11.6	10	18	52	5	10	5
			*11.6–12.6	8	43	33	3	6	7
			*12.6–13.8	5	15	32	17	16	15
			*13.8–14.9	13	7	39	13	16	12
			Mean	11	18	38	9	15	9

TL 61 SE 16 6841 1314 Humphrey's Farm **Block E**

Surface level (+58.8 m) +193 ft
Water not struck
152 mm percussion
June 1977

Waste 16.0
Bedrock 1.0+

LOG

Geological classification	Lithology	Thickness m	Depth m
	Soil	0.4	0.4
Boulder Clay	Clay, brown with chalk and flint pebbles, becoming grey with additional pebbles of shale and limestone below 8.3 m	14.0	14.4
	'Very clayey' sand, with some pebbles Gravel: coarse and fine, angular and well-rounded flint with quartz and quartzite Sand: medium with some coarse and fine, light grey	1.6	16.0
London Clay	Clay, stiff, olive-grey	1.0+	17.0

TL 61 SE 17 6818 1037 Woodside **Block E**

Surface level (+48.5 m) +159 ft
Water not struck
152 mm percussion
May 1977

Waste 8.7
Bedrock 0.5+

LOG

Geological classification	Lithology	Thickness m	Depth m
	Soil	0.4	0.4
Boulder Clay	Clay, brown with chalk and flint pebbles, mottled to light grey below 1.5 m	3.6	4.0
	Clay, brown with sandy seams	2.8	6.8
Glacial Sand and Gravel	Gravel Gravel: coarse and fine, angular and well-rounded flint with rounded quartz and quartzite Sand: medium and coarse with fine	1.9	8.7
London Clay	Clay, stiff, light brown	0.5+	9.2

GRADING

Mean for deposit *percentages*			Depth below surface (m)	*percentages*					
Fines	Sand	Gravel		Fines	Sand			Gravel	
				$-\frac{1}{16}$	$+\frac{1}{16}-\frac{1}{4}$	$+\frac{1}{4}-1$	+1–4	+4–16	+16–64
1	29	70	*6.8–7.8	1	7	26	9	31	26
			*7.8–8.7	2	0	2	12	31	53
			Mean	1	4	15	10	31	39

TL 61 SE 18 6939 1477 Fitzandrew's Farm **Block D**

Surface level (+47.9 m) +157 ft
Water struck at +43.4 m
152 mm percussion
June 1977

Overburden 3.3
Mineral 7.4
Bedrock 1.1+

LOG

Geological classification	Lithology	Thickness m	Depth m
	Soil and subsoil	0.3	0.3
Boulder Clay	Clay, very sandy, orange-brown with flint, quartzite and quartz pebbles	3.0	3.3
Glacial Sand and Gravel	Gravel, 'very clayey' above 4.5 m Gravel: fine and coarse, angular to subangular with well-rounded flint, rounded to subrounded quartzite, some quartz and sandstone Sand: medium with coarse and fine, quartz with flint, orange-grey	7.4	10.7
London Clay	Clay, stiff, olive-grey, top 0.7 m weathered to brown	1.1+	11.8

GRADING

Mean for deposit *percentages*			Depth below surface (m)	*percentages*					
Fines	Sand	Gravel		Fines	Sand			Gravel	
				$-\frac{1}{16}$	$+\frac{1}{16}-\frac{1}{4}$	$+\frac{1}{4}-1$	+1–4	+4–16	+16–64
7	40	53	3.3–4.5	32	30	17	4	12	5
			*4.5–5.6	2	15	26	5	25	27
			*5.6–6.7	3	5	13	6	31	42
			*6.7–7.8	1	4	24	16	25	30
			*7.8–9.2	1	5	25	13	31	25
			*9.2–10.7	5	3	13	16	33	30
			Mean	7	10	20	10	27	26

COMPOSITION

Depth below surface (m)	Percentages by weight in gravel fraction						
	Angular flint	Well-rounded flint	Quartzite	Quartz	Sandstone	Chalk	Others
3.3–10.7	51	22	21	4	2	0	0

TL 61 SE 19 6981 1239 Ball's Farm **Block E**

Surface level (+51.8 m) +170 ft
Water struck at +43.8 m
152 mm percussion
June 1977

Overburden 5.0
Mineral 6.8
Bedrock 0.7+

LOG

Geological classification	Lithology	Thickness m	Depth m
	Soil and subsoil	0.4	0.4
Boulder Clay	Clay, orange-brown with chalk and flint pebbles, mottled to light grey below 1.0 m	4.1	4.5
Glacial Sand and Gravel	Clay, very sandy, orange-brown with chalk and flint pebbles	0.5	5.0
	Gravel, with 'clayey' seams from 5.6 to 7.8 m Gravel: coarse and fine, angular to subangular with well-rounded flint, rounded to subrounded quartzite, quartz and some sandstone Sand: medium and coarse with fine, orange-grey	6.8	11.8
London Clay	Clay, stiff, olive-grey, top 0.4 m weathered to brown	0.7+	12.5

GRADING

Mean for deposit *percentages* Fines	Sand	Gravel	Depth below surface (m)	*percentages* Fines $-\frac{1}{16}$	Sand $+\frac{1}{16}-\frac{1}{4}$	$+\frac{1}{4}-1$	$+1-4$	Gravel $+4-16$	$+16-64$
4	33	63	5.0–5.6	29	8	20	8	23	12
			*5.6–6.8	3	5	18	9	30	35
			*6.8–8.0	2	3	17	12	32	34
			*8.0–9.0	4	3	10	10	31	42
			*9.0–11.8	1	2	12	20	34	31
			Mean	4	4	15	14	31	32

COMPOSITION

Depth below surface (m)	Percentages by weight in gravel fraction: Angular flint	Well-rounded flint	Quartzite	Quartz	Sandstone	Chalk	Others
5.0–11.8	55	13	23	6	2	0	1

TL 61 SE 20 6962 1141 Partridgegreen Cottages **Block E**

Surface level (+60.4 m) +198 ft
Water not struck
152 mm percussion
June 1977

Waste 18.9+

LOG

Geological classification	Lithology	Thickness m	Depth m
	Soil	0.2	0.2
Boulder Clay	Clay, orange-brown with flints	1.3	1.5
	Clay, orange-brown with chalk and flint pebbles, mottled to light grey from 3.0 m, becoming olive-grey with additional pebbles of shale, siltstone and sandstone below 10.2 m	17.4+	18.9

TL 61 SE 21 6966 1028 Stacey's Farm **Block E**

Surface level (+53.6 m) +176 ft
Water not struck
152 mm percussion
June 1977

Overburden 7.7
Mineral 9.9
Bedrock 0.9+

LOG

Geological classification	Lithology	Thickness m	Depth m
	Soil	0.3	0.3
Boulder Clay	Clay, brown with flints	1.2	1.5
	Clay, brown with chalk and flint pebbles. Passing down into clay, sandy, brown with flints below 5.5 m	6.2	7.7
Glacial Sand and Gravel	Gravel Gravel: fine and coarse, angular to subangular with well-rounded flint, rounded to subrounded quartzite, some quartz and sandstone Sand: medium and coarse with fine, quartz with flint, orange-grey	9.9	17.6
London Clay	Clay, stiff, brown	0.9+	18.5

GRADING

Mean for deposit *percentages*			Depth below surface (m)	*percentages*					
Fines	Sand	Gravel		Fines	Sand			Gravel	
				$-\frac{1}{16}$	$+\frac{1}{16}-\frac{1}{4}$	$+\frac{1}{4}-1$	+1–4	+4–16	+16–64
5	36	59	*7.7–8.9	2	4	17	7	32	38
			*8.9–10.0	6	2	11	7	42	32
			*10.0–11.1	7	7	22	11	31	22
			*11.1–12.7	2	6	25	21	25	21
			*12.7–13.7	5	5	19	15	31	25
			*13.7–15.2	1	1	11	17	41	29
			*15.2–16.2	11	1	13	10	38	27
			*16.2–17.6	7	4	24	19	23	23
			Mean	5	4	18	14	32	27

COMPOSITION

Depth below surface (m)	Percentages by weight in gravel fraction						
	Angular flint	Well-rounded flint	Quartzite	Quartz	Sandstone	Chalk	Others
7.7–17.6	65	15	14	4	2	0	0

APPENDIX G

List of workings

In 1978 the two sand and gravel pits listed were known to be operational. All the worked-out areas are shown on the map accompanying the report.

Location	*Grid reference*	*Deposit*
Chignall Hall	660 104	Glacial Sand and Gravel
Margret at Woods	686 119	Glacial Sand and Gravel

APPENDIX H

CONVERSION TABLE, METRES TO FEET (to nearest 0.5 ft)

m	ft	m	ft	m	ft	m	ft	m	ft
0.1	0.5	6.1	20	12.1	39.5	18.1	59.5	24.1	79
0.2	0.5	6.2	20.5	12.2	40	18.2	59.5	24.2	79.5
0.3	1	6.3	20.5	12.3	40.5	18.3	60	24.3	79.5
0.4	1.5	6.4	21	12.4	40.5	18.4	60.5	24.4	80
0.5	1.5	6.5	21.5	12.5	41	18.5	60.5	24.5	80.5
0.6	2	6.6	21.5	12.6	41.5	18.6	61	24.6	80.5
0.7	2.5	6.7	22	12.7	41.5	18.7	61.5	24.7	81
0.8	2.5	6.8	22.5	12.8	42	18.8	61.5	24.8	81.5
0.9	3	6.9	22.5	12.9	42.5	18.9	62	24.9	81.5
1.0	3.5	7.0	23	13.0	42.5	19.0	62.5	25.0	82
1.1	3.5	7.1	23.5	13.1	43	19.1	62.5	25.1	82.5
1.2	4	7.2	23.5	13.2	43.5	19.2	63	25.2	82.5
1.3	4.5	7.3	24	13.3	43.5	19.3	63.5	25.3	83
1.4	4.5	7.4	24.5	13 4	44	19.4	63.5	25.4	83.5
1.5	5	7.5	24.5	13.5	44.5	19.5	64	25.5	83.5
1.6	5	7.6	25	13.6	44.5	19.6	64.5	25.6	84
1.7	5.5	7.7	25.5	13.7	45	19.7	64.5	25.7	84.5
1.8	6	7.8	25.5	13.8	45.5	19.8	65	25.8	84.5
1.9	6	7.9	26	13.9	45.5	19.9	65.5	25.9	85
2.0	6.5	8.0	26	14.0	46	20.0	65.5	26.0	85.5
2.1	7	8.1	26.5	14.1	46.5	20.1	66	26.1	85.5
2.2	7	8.2	27	14.2	46.5	20.2	66.5	26.2	86
2.3	7.5	8.3	27	14.3	47	20.3	66.5	26.3	86.5
2.4	8	8.4	27.5	14.4	47	20.4	67	26.4	86.5
2.5	8	8.5	28	14.5	47.5	20.5	67.5	26.5	87
2.6	8.5	8.6	28	14.6	48	20.6	67.5	26.6	87.5
2.7	9	8.7	28.5	14.7	48	20.7	68	26.7	87.5
2.8	9	8.8	29	14.8	48.5	20.8	68	26.8	88
2.9	9.5	8.9	29	14.9	49	20.9	68.5	26.9	88.5
3.0	10	9.0	29.5	15.0	49	21.0	69	27.0	88.5
3.1	10	9.1	30	15.1	49.5	21.1	69	27.1	89
3.2	10.5	9.2	30	15.2	50	21.2	69.5	27.2	89
3.3	11	9.3	30.5	15.3	50	21.3	70	27.3	89.5
3.4	11	9.4	31	15.4	50.5	21.4	70	27.4	90
3.5	11.5	9.5	31	15.5	51	21.5	70.5	27.5	90
3.6	12	9.6	31.5	15.6	51	21.6	71	27.6	90.5
3.7	12	9.7	32	15.7	51.5	21.7	71	27.7	91
3.8	12.5	9.8	32	15.8	52	21.8	71.5	27.8	91
3.9	13	9.9	32.5	15.9	52	21.9	72	27.9	91.5
4.0	13	10.0	33	16.0	52.5	22.0	72	28.0	92
4.1	13.5	10.1	33	16.1	53	22.1	72.5	28.1	92
4.2	14	10.2	33.5	16.2	53	22.2	73	28.2	92.5
4.3	14	10.3	34	16.3	53.5	22.3	73	28.3	93
4.4	14.5	10.4	34	16.4	54	22.4	73.5	28.4	93
4.5	15	10.5	34.5	16.5	54	22.5	74	28.5	93.5
4.6	15	10.6	35	16.6	54.5	22.6	74	28.6	94
4.7	15.5	10.7	35	16.7	55	22.7	74.5	28.7	94
4.8	15.5	10.8	35.5	16.8	55	22.8	75	28.8	94.5
4.9	16	10.9	36	16.9	55.5	22.9	75	28.9	95
5.0	16.5	11.0	36	17.0	56	23.0	75.5	29.0	95
5.1	17	11.1	36.5	17.1	56	23.1	76	29.1	95.5
5.2	17	11.2	36.5	17.2	56.5	23.2	76	29.2	96
5.3	17.5	11.3	37	17.3	57	23.3	76.5	29.3	96
5.4	17.5	11.4	37.5	17.4	57	23.4	77	29.4	96.5
5.5	18	11.5	37.5	17.5	57.5	23.5	77	29.5	97
5.6	18.5	11.6	38	17.6	57.5	23.6	77.5	29.6	97
5.7	18.5	11.7	38.5	17.7	58	23.7	78	29.7	97.5
5.8	19	11.8	38.5	17.8	58.5	23.8	78	29.8	98
5.9	19.5	11.9	39	17.9	58.5	23.9	78.5	29.9	98
6.0	19.5	12.0	39.5	18.0	59	24.0	78.5	30.0	98.5

REFERENCES

ALLEN V. T. 1936. Terminology of medium grained sediments. *Rep. Natl Res. Counc. Washington 1935–1936, App. 1, Rep. Comm. Sedimentation,* pp. 18–47.

ARCHER, A. A. 1969. Background and problems of an assessment of sand and gravel resources in the United Kingdom. *Proc. 9th Commonw. Min. Metall. Congr., 1969.* Vol. 2, Mining and Petroleum Geology, pp. 495–508.

—— 1970a. Standardisation of the size classification of naturally occurring particles. *Géotechnique,* Vol. 20, pp. 103–107.

—— 1970b. Making the most of metrication. *Quarry Managers' J.,* Vol. 54, No. 6, pp. 223–227.

ATTERBERG, A. 1905. Die rationelle Klassifikation der Sande und Kiese. *Chem. Z.,* Vol. 29, pp. 195–198.

BRITISH STANDARD 1377. 1967. *Methods of testing soils for civil engineering purposes.* (London: British Standards Institution.) 233 pp.

BUREAU OF MINES AND GEOLOGICAL SURVEY. 1948. *Mineral Resources of the United States* (Washington D. C.: Public Affairs Press.) pp. 14–17.

EATON, C. H. 1973. The sand and gravel resources of the country around Terling, Essex: Description of 1: 25 000 resource sheet TL 71. *Rep. Inst. Geol. Sci.,* No. 73/5, 120 pp.

HARRIS, P. M., THURRELL, R. G., HEALING, R. A. and ARCHER, A. A. 1974. Aggregates in Britain. *Proc. R. Soc. London,* Ser. A. 339, pp. 329–353.

HESTER, S. W. 1965. Stratigraphy and palaeogeography of the Woolwich and Reading Beds. *Bull. Geol. Surv. G.B.,* No. 23, pp. 117–137.

HOPSON, P. M. 1979. The sand and gravel resources of the country north of Harlow, Essex: Description of 1:25 000 resource sheet TL 41. *Miner. Assess. Rep. Inst. Geol. Sci.* No. 46.

LANE, E. W. and others, 1947. Report of the sub-committee on sediment terminology. *Trans. Am. Geophys. Union,* Vol. 28, pp. 936–938.

PETTIJOHN, F. J. 1957. *Sedimentary Rocks.* 2nd Edition (London: Harper and Row.)

ROSE, J., ALLEN, P. and HEY, R. W. 1976. Middle Pleistocene stratigraphy in southern East Anglia. *Nature,* Vol. 263, pp. 492–494.

— — 1977. Middle Pleistocene stratigraphy in south-east Suffolk. *J. Geol. Soc. London,* Vol. 133, pp. 83–102.

THURRELL, R. G. 1971. The assessment of mineral resources with particular reference to sand and gravel. *Quarry Managers' J.,* Vol. 55, pp. 19–25.

TWENHOFEL, W. H. 1937. Terminology of the fine-grained mechanical sediments. *Rep. Natl Res. Counc. Washington, 1936–1937. App. 1, Rep. Comm. Sedimentation,* pp. 81–104.

UDDEN. J. A. 1914. Mechanical composition of elastic sediments. *Bull. Geol. Soc. Am.,* Vol. 25, pp. 655–744.

WENTWORTH, C. K. 1922. A scale of grade and class terms for clastic sediments. *J. Geol.,* Vol. 30, pp. 377–392.

—— 1935. The terminology of coarse sediments. *Bull. Natl Res. Counc. Washington,* No. 98, pp. 225–246.

WHITAKER, W., PENNING, W. H., DALTON, W. H. and BENNETT, F. J. 1878. The geology of the N.W. part of Essex and the N.E. part of Herts. with parts of Cambridgeshire and Suffolk. *Mem. Geol. Surv. G.B.*

WILLMAN, H. B. 1942. Geology and mineral resources of the Marseilles, Ottawa and Streator quadrangles. *Bull. Illinois State Geol. Surv., No. 66,* pp. 343–344.

WOOLDRIDGE, S. W. and HENDERSON, H. C. K. 1955. Some aspects of the physiography of the eastern part of the London Basin. *Trans. I.B.G.,* Vol. 21, pp. 19–31.

The following reports of the Institute relate particularly to bulk mineral resources

Reports of the Institute of Geological Sciences

Assessment of British Sand and Gravel Resources

1 The sand and gravel resources of the country south-east of Norwich, Norfolk: Resource sheet TG 20. E. F. P. Nickless.
Report 71/20 ISBN 0 11 880216 £1.15

2 The sand and gravel resources of the country around Witham, Essex: Resource sheet TL 81. H. J. E. Haggard.
Report 72/6 ISBN 0 11 880588 6 £1.20

3 The sand and gravel resources of the area south and west of Woodbridge, Suffolk: Resource sheet TM 24. R. Allender and S. E. Hollyer.
Report 72/9 ISBN 0 11 880596 7 £1.70

4 The sand and gravel resources of the country around Maldon, Essex: Resource sheet TL 80. J. D. Ambrose.
Report 73/1 ISBN 0 11 880600 9 £1.20

5 The sand and gravel resources of the country around Hethersett, Norfolk: Resource sheet TG 10. E. F. P. Nickless.
Report 73/4 ISBN 0 11 880606 8 £1.60

6 The sand and gravel resources of the country around Terling, Essex: Resource sheet TL 71. C. H. Eaton.
Report 73/5 ISBN 0 11 880608 4 £1.20

7 The sand and gravel resources of the country around Layer Breton and Tolleshunt D'Arcy, Essex: Resource sheet TL 91 and part of TL 90. J. D. Ambrose.
Report 73/8 ISBN 0 11 990614 9 £1.30

8 The sand and gravel resources of the country around Shotley and Felixstowe, Suffolk: Resource sheet TM 23. R. Allender and S. E. Hollyer.
Report 73/13 ISBN 0 11 880625 4 £1.60

9 The sand and gravel resources of the country around Attlebridge, Norfolk: Resource sheet TG 11. E. F. P. Nickless.
Report 73/15 ISBN 0 11 880658 0 £1.85

10 The sand and gravel resources of the country west of Colchester, Essex: Resource sheet TL 92. J. D. Ambrose.
Report 74/6 ISBN 0 11 880671 8 £1.45

11 The sand and gravel resources of the country around Tattingstone, Suffolk: Resource sheet TM 13. S. E. Hollyer.
Report 74/9 ISBN 0 11 880675 0 £1.95

12 The sand and gravel resources of the country around Gerrards Cross, Buckinghamshire: Resource sheet SU 99, TQ 08 and TQ 09. H. C. Squirrell.
Report 74/14 ISBN 0 11 880710 2 £2.20

Mineral Assessment Reports

13 The sand and gravel resources of the country east of Chelmsford, Essex: Resource sheet TL 70. M. R. Clarke.
ISBN 0 11 880744 7 £3.50

14 The sand and gravel resources of the country east of Colchester, Essex: Resource sheet TM 02. J. D. Ambrose.
ISBN 0 11 880745 5 £3.25

15 The sand and gravel resources of the country around Newton on Trent, Lincolnshire: Resource sheet SK 87. D. Price.
ISBN 0 11 880746 3 £3.00

16 The sand and gravel resources of the country around Braintree, Essex: Resource sheet TL 72. M. R. Clarke.
ISBN 0 11 880747 1 £3.50

17 The sand and gravel resources of the country around Besthorpe, Nottinghamshire: Resource sheet SK 86 and part of SK 76. J. R. Gozzard.
ISBN 0 11 880748 X £3.00

18 The sand and gravel resources of the Thames Valley, the country around Cricklade, Wiltshire: Resource sheet SU 09/19 and parts of SP 00/10. P. Robson.
ISBN 0 11 880749 8 £3.00

19 The sand and gravel resources of the country south of Gainsborough, Lincolnshire: Resource sheet SK 88 and part of SK 78. J. H. Lovell.
ISBN 0 11 880750 1 £2.50

20 The sand and gravel resources of the country east of Newark upon Trent, Nottinghamshire: Resource sheet SK 85. J. R. Gozzard.
ISBN 0 11 880751 X £2.75

21 The sand and gravel resources of the Thames and Kennet Valleys, the country around Pangbourne, Berkshire: Resource sheet SU 67. H. C. Squirrell.
ISBN 0 11 880752 8 £3.25

22 The sand and gravel resources of the country north-west of Scunthorpe, Humberside: Resource sheet SE 81. J. W. C. James.
ISBN 0 11 880753 6 £3.00

23 The sand and gravel resources of the Thames Valley, the country between Lechlade and Standlake: Resource sheet SP 30 and parts of SP 20, SU 29 and SU 39. P. Robson.
ISBN 0 11 881252 1 £7.25

24 The sand and gravel resources of the country around Aldermaston, Berkshire: Resource sheet SU 56 and SU 66. H. C. Squirrell.
ISBN 0 11 881253 X £5.00

25 The celestite resources of the area north-east of Bristol: Resource sheet ST 68 and parts of ST 59, 69, 79, 58, 78, 68 and 77. E. F. P. Nickless, S. J. Booth and P. N. Mosley.
ISBN 0 11 881262 9 £5.00

26 The limestone and dolomite resources of the country around Monyash, Derbyshire: Resource sheet SK 16. F. C. Cox and D. McC. Bridge.
ISBN 0 11 881263 7 £7.00

27 The sand and gravel resources of the country west and south of Lincoln, Lincolnshire: Resource sheets SK 95, SK 96 and SK 97. I. Jackson.
ISBN 0 11 884003 7 £6.00

28 The sand and gravel resources of the country around Eynsham, Oxfordshire: Resource sheet SP 40 and part of SP 41. W. J. R. Harries.
ISBN 0 11 884012 6 £3.00

29 The sand and gravel resources of the country south-west of Scunthorpe, Humberside: Resource sheet SE 80. J. H. Lovell.
ISBN 0 11 884013 4 £3.50

30 Procedure for the assessment of limestone resources. F. C. Cox, D. McC. Bridge and J. H. Hull.
ISBN 0 11 884030 4 £1.25

31 The sand and gravel resources of the country west of Newark upon Trent, Nottinghamshire. Resource sheet SK 75. D. Price and P. J. Rogers.
ISBN 0 11 884031 2 £3.50

32 The sand and gravel resources of the country around Sonning and Henley: Resource sheet SU 77 and SU 78. H. C. Squirrell.
ISBN 0 11 884032 0 £5.25

33 The sand and gravel resources of the country north of Gainsborough: Resource sheet SK 89. J. R. Gozzard and D. Price.
ISBN 0 11 884033 9 £4.50

34 The sand and gravel resources of the Dengie Peninsula, Essex: Resource sheet TL 90, etc. M. B. Simmons.
ISBN 0 11 884081 9 £5.00

35 The sand and gravel resources of the country around Darvel: Resource sheet NS 53, 63, etc. E. F. P. Nickless, A. M. Aitken and A. A. McMillan.
ISBN 0 11 884082 7 £7.00

36 The sand and gravel resources of the country around Southend-on-Sea, Essex: Resource sheets TQ 78/79 etc. S. E. Hollyer and M. B. Simmons.
ISBN 0 11 884083 5 £7.50

37 The sand and gravel resources of the country around Bawtry, South Yorkshire: Resource sheet SK 69. A. R. Clayton.
ISBN 0 11 884053 3 £5.75

38 The sand and gravel resources of the country around Abingdon, Oxfordshire: Resource sheet SU 49, 59, SP 40, 50. C. E. Corser.
ISBN 0 11 884084 5 £5.50

39 The sand and gravel resources of the Blackwater Valley (Aldershot) area: Resource sheet SU 85, 86, parts SU 84, 94, 95, 96. M. R. Clarke, A. J. Dixon and M. Kubala.
ISBN 0 11 884085 1 £7.00

40 The sand and gravel resources of the country west of Darlington, County Durham: Resource sheet NZ 11, 21. A. Smith.
ISBN 0 11 884086 X £5.00

41 The sand and gravel resources of the country around Garmouth, Grampian Region: Resource sheet NJ 36. A. M. Aitken, J. W. Merritt and A. J. Shaw.
ISBN 0 11 884090 8 £8.75

42 The sand and gravel resources of the country around Maidenhead and Marlow: Resource sheet SU 88, parts SU 87, 97, 98. P. N. Dunkley.
ISBN 0 11 884091 6 £5.00

43 The sand and gravel resources of the country around Misterton, Nottinghamshire: Resource sheet SK 79. D. Thomas and D. Price.
ISBN 0 11 884092 4 £5.25

44 The sand and gravel resources of the country around Sedgefield, Durham: resource sheet NZ 32. M. D. A. Samuel.
ISBN 0 11 884093 2 £5.75

45 The sand and gravel resources of the country around Brampton, Cumbria: Resource sheet NY 55, part 56. I. Jackson.
ISBN 0 11 884094 0 £6.75

46 The sand and gravel resources of the country around Harlow, Essex: Resource sheet TL 41. P. M. Hopson.
ISBN 0 11 884107 6 *not yet priced*

47 The limestone and dolomite resources of the country around Wirksworth, Derbyshire: Resource sheet SK 25, part 35. F. C. Cox and D. J. Harrison.
ISBN 0 11 884108 4 £9.75

48 The sand and gravel resources of the Loddon valley area: Resource sheets SU 75, 76, parts 64, 65, 66 and 74. M. R. Clarke, E. J. Raynor and R. A. Sobey.
ISBN 0 11 884109 2 £8.75

49 The sand and gravel resources of the country around Lanark, Strathclyde Region: Resource sheet NS 94, part 84. J. L. Laxton and E. F. P. Nickless.
ISBN 0 11 884112 2 £11.00

50 The sand and gravel resources of the country around Fordingbridge, Hampshire: Resource sheet SU 11 and parts of SU 00, 01, 10, 20 and 21. M. Kubala.
ISBN 0 11 884111 4 £7.75

51 The sand and gravel resources of the country north of Bournemouth, Dorset: Resource sheets SU 00, 10, 20, SZ 09, 19 and 29, M. R. Clarke.
ISBN 0 11 884110 6 *not yet priced*

52 The sand and gravel resources of the country between Hatfield Heath and Great Waltham, Essex: Resource sheet TL 51 and 61. R. J. Marks.
ISBN 0 11 884113 0 £8.00

Reports of the Institute of Geological Sciences

Other Reports

69/9 Sand and gravel resources of the inner Moray Firth. A. L. Harris and J. D. Peacock.
ISBN 0 11 880106 6 35p

70/4 Sands and gravels of the southern counties of Scotland. G. A. Goodlet.
ISBN 0 11 880105 8 90p

72/8 The use and resources of moulding sand in Northern Ireland. R. A. Old.
ISBN 0 11 881594 0 30p

73/9 The superficial deposits of the Firth of Clyde and its sea lochs. C. E. Deegan, R. Kirby, I. Rae and R. Floyd.
ISBN 0 11 880617 3 95p

77/1 Sources of aggregate in Northern Ireland (2nd edition). I. B. Cameron.
ISBN 0 11 881279 3 70p

77/2 Sand and gravel resources of the Grampian region. J. D. Peacock and others.
ISBN 0 11 881282 3 80p

77/5 Sand and gravel resources of the Fife Region. M. A. E. Browne.
ISBN 0 11 884004 5 60p

77/6 Sand and gravel resources of the Tayside Region. I. B. Paterson.
ISBN 0 11 884008 8 £1.40

77/8 Sand and gravel resources of the Strathclyde Region. I. B. Cameron and others.
ISBN 0 11 884028 2 £2.50

77/9 Sand and gravel resources of the Central Region, Scotland. M. A. E. Browne.
ISBN 0 11 884016 9 £1.35

77/19 Sand and gravel resources of the Borders region, Scotland. A. D. McAdam.
ISBN 0 11 884025 8 £1.00

77/22 Sand and gravel resources of the Dumfries and Galloway Region of Scotland. I. B. Cameron.
ISBN 0 11 884025 8 £1.20

78/1 Sand and gravels of the Lothian Region of Scotland. A. D. McAdam.
ISBN 0 11 884042 8 £1.00

78/8 Sand and gravel resources of the Highland Region. W. Mykura, D. L. Ross and F. May.
ISBN 0 11 884050 9 £3.00

Dd 696798 K8

Typeset for the Institute of Geological Sciences by John Wright and Sons Ltd, Bristol

Printed in England for Her Majesty's Stationery Office by Commercial Colour Press, London E7